T0269307

WAFL 2014

Proceedings of the 6th International Conference on the Assessment of Animal Welfare at Farm and Group Level

WAFL 2014

Clermont-Ferrand, France
September 3-5, 2014

edited by:
Luc Mounier and Isabelle Veissier

Proceedings publication and Abstract Submission System (OASES) by

Wageningen Academic
P u b l i s h e r s

EAN: 9789086862474
e-EAN: 9789086867981
ISBN: 978-90-8686-247-4
e-ISBN: 978-90-8686-798-1
DOI: 10.3920/978-90-8686-798-1

First published, 2014

© Wageningen Academic Publishers
The Netherlands, 2014

This work is subject to copyright. All rights are reserved, whether the whole or part of the material is concerned. Nothing from this publication may be translated, reproduced, stored in a computerised system or published in any form or in any manner, including electronic, mechanical, reprographic or photographic, without prior written permission from the publisher: Wageningen Academic Publishers P.O. Box 220 6700 AE Wageningen The Netherlands www.WageningenAcademic.com copyright@WageningenAcademic.com

The individual contributions in this publication and any liabilities arising from them remain the responsibility of the authors.

The publisher is not responsible for possible damages, which could be a result of content derived from this publication.

Foreword

WAFL is an international scientific conference on the assessment of animal welfare at farm or group level. Issues addressed during a WAFL conference can be:
- Welfare criteria,
  Animal welfare comprises many aspects. The choice of welfare criteria to be covered is crucial before any assessment method is developed;
- Welfare indicators and methods to analyse them,
  Valid methods to assess animal welfare in experiments may not be applicable in commercial settings. Methods adapted for monitoring animal welfare at group or farm level must be developed and validated. They are usually simpler and less time consuming. Their quality and conditions of use should be specified;
- Automation of measures,
  The automation of measures offers new use of welfare indicators in on-farm welfare assessments. Devices for automatic recording of data and models to interpret the data into meaningful welfare information are developed;
- Ways to deal with a large amount of data from various welfare measures in order to assess the overall welfare of animals.
  This questioned not only animal sciences but also ethics;
- Rationales underlying epidemiological studies and risk models.
  Epidemiological studies can help us identify conditions that result in poor vs. good welfare. The principles of such studies could be mode widely spread among animal welfare scientists;
- Consultation processes.
  The choice of welfare criteria and the design of an overall welfare assessment cannot entirely be based on science. It is generally done after a consultation process. Ways to organise this consultation can be formalised;
- Implementation in practice.
  How welfare measures are used in welfare plans to certify farms (or other types of animal units) or to monitor progresses should be shared.
- Communication of animal welfare assessment results.
  Animal welfare may mean different things to different parties such as farmers, laypeople and experts.
- And other topics related to animal welfare assessment in practice.

WAFL2014 is the 6th edition of the conference. WAFL is typically a multidisplinary conference. It is also largely open to a non-scientific audience. We believe a transdisciplinary approach is requested to go between and beyond disciplines and engage with society at large in order to stimulate innovations that can improve animal welfare.

Isabelle Veissier and Luc Mounier

Acknowledgements

Scientific committee

The members of the scientific committee are warmly thanked for their essential role in designing the WAFL2014 program and selecting proposals. We are grateful to them for being so quick at responding to our requests and for the high quality of their inputs.

Isabelle Veissier (Chair)
Luc Mounier (Co-chair)
Valérie Courboulay
Björn Forkman
Florence Kling-Eveillard
Stella Maris Huertas Canen
Marie-Christine Meunier-Salaün
Mara Miele
Jeffrey Rushen
Marek Spinka
Hans Spoolder
Claudia Terlouw

Organizing committee

The members of the organizing committee are warmly thanked for their personal contribution to make WAFL2014 a high quality event.

Luc Mounier (Chair)
Stephane Andanson (Co-chair)
Nicolas Barrière
Anne-Marie Chanel
Christiane Espinasse

The logo was designed by Jimmylie Fontes

Thank you to our sponsors

Silver Sponsors

Bronze Sponsors

Scientific abstracts

Keynotes

Session 1. Ethical issues in relation to animal welfare assessment

Session 2. Consultation processes

Session 3. Welfare indicators and methods to analyse them

Session 4. Automation of welfare measurements

Session 5. Statistical methods to deal with large amounts of data from various welfare measures

Session 6. Rationales underlying epidemiological studies and risk models for animal welfare

Session 7. Training assessors

Session 8. Implementation of animal welfare assessment, to certify farms or for other purposes

Session 9. Cost-benefit analyses of implementation of welfare assessment systems

Session 10. Communication of animal welfare assessment results

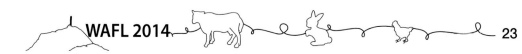

Training assessors

Implementation of animal welfare assessment, to certify farms or for other purposes

Cost-benefit analyses of implementation of welfare assessment systems

Communication of animal welfare assessment results

Much still to do: Ruth Harrison, 'Animal Machines' and the state of animal welfare science

M.S. Dawkins

University of Oxford, Zoology, South Parks Road, Oxford OX1 3PS, United Kingdom;
marian.dawkin@zoo.ox.ac.uk

A striking features of Ruth Harrison's book Animal Machines is the breadth of its coverage and the many different arguments that she used to make her points. Ruth Harrison did not see the welfare of farm animals as an isolated issue or as a narrow concern to a small group of people. Rather, she saw that the way farm animals are treated has implications for human society, through its impact on human health, human well-being, the environment and the economy. She asked Rachel Carson, author of Silent Spring, to write the foreword to Animal Machines because she could see the connections between the ethical treatment of the environment and the ethical treatment of animals. Today, 50 years later, we still have much to learn from Ruth Harrison's broad-based approach to animal welfare. All too often animal welfare is studied in isolation from its implications for other disciplines and 'solutions' put forward that do not adequately consider the economic, environmental and other issues that may be crucial in deciding whether or not they are implemented. I shall start by setting animal welfare in the context of current concerns about food security and the need to mitigate the effects of climate change. I shall argue that the pressures for agriculture to become more efficient and more 'sustainably intensive' pose a threat to animal welfare unless we can, as Ruth Harrison did, argue the case for animal welfare on many different grounds, including economic ones. This is the only way that animal welfare will be taken seriously in a world increasingly concerned with human survival. I will then look at a number of recent developments in such as the growth of precision farming and the use of technology to monitor farm animal welfare and show that, properly handled, they hold out the promise for improvements in animal welfare. However, these improvements in animal welfare will not happen by themselves, nor will they happen if 'good welfare' is promoted in isolation from other factors such as what is good for people and what enables farmers to make a living. The agenda that Ruth Harrison set out in Animal Machines is still a very good guide for the future of animal welfare.

Play as an indicator of good welfare

S. Held

Animal Welfare and Behaviour Group, School of Veterinary Sciences, University of Bristol, United Kingdom; suzanne.held@bristol.ac.uk

Play behaviour commonly disappears under fitness challenge, and is thought to be accompanied by a pleasurable emotional experience. It has therefore long been suggested as a welfare indicator, particularly of good welfare. But animal play is a tricky behavioural phenomenon. It is highly variable between and within species, and its biological functions are not yet fully understood. This talk therefore reviews evidence for and against play always indicating good welfare in farm animals; and it considers biological and practical factors affecting its usefulness as an indicator at group level.

How epidemiology can help researches in animal welfare

N. Bareille[1] and L. Mounier[3,4]
[1]INRA, UMR1300, Biology, Epidemiology and Risk Analysis in Animal Health, CS 40706, 44307 Nantes, France, [2]LUNAM Université, Oniris, Nantes-Atlantic College of Veterinary Medicine, Food Sciences and Engineering, UMR BioEpAR, 44307 Nantes, France, [3]Université de Lyon, VetAgro Sup, UMR 1213 Herbivores, 69280 Marcy L'Etoile, France, [4]INRA, UMR1213 Herbivores, 63122 Saint-Genes-Champanelle, France; nathalie.bareille@oniris-nantes.fr

After years of experimental contributions, animal welfare researches expand to field approaches in order to learn what really occurs in farms. Concepts of epidemiology can be useful to produce robust results. Indeed, veterinary epidemiology consists in the study of the health status of animal populations, i.e. the description of disease frequency and the determination of risk factors for their occurrence. The same approach can be applied to animal welfare with assessment of welfare and identification of risk factors. We will illustrate how the methods used in veterinary epidemiology can be used to study animal welfare. The first aim of epidemiological studies is to describe the health status of a given animal population. This description is seldom done on the whole population due to the lack of a mandatory data recording system. Therefore, a random sampling strategy must be applied to obtain a sample representative of the targeted population. Due to differences in duration among diseases, the disease frequency is expressed as prevalence for long lasting disorders e.g. lameness, or as cumulative incidence for short lasting disorders e.g. mortality. The Welfare Quality® protocol clearly uses these two descriptors: prevalence, when data comes from the animal inspection on the day of the farm visit (most Welfare quality® measures), and incidence, for retrospective data collection. The main problem in descriptive epidemiology is the definition of a diseased animal and the value of the measurement method used to determine the health status of a given animal. For infectious diseases, the diagnostic is based on the identification of the pathogen. For the non-infectious diseases, such as lameness ,a variety of definitions (lame animal treated, animals with abnormal leg position …) and tools are used without obvious reference method. This leads to difficulties to compare studies and situations. This problem is crucial in animal welfare studies where gold standards are rarely available. The methods used in the Welfare Quality® protocols may be used as standard if they become widely used. Even without gold standards, descriptive surveys can provide a benchmark for farmers and advisors involved in an improvement process. The second aim of epidemiological studies is to identify the determinants of the disease occurrence, i.e. risk factors. The epidemiologist then analyses the links between the disease occurrence and the exposure of the animals to potential risk factors (individual characteristics, management practices …). To this aim, he (she) implements a statistical model to find out associations between these two events. Nevertheless, the causal relationship between exposure and disease occurrence cannot be demonstrated because of potential bias, confounding and other pitfalls. Nine conditions have been proposed to help assess causality in epidemiological studies. The main ones are the plausibility of the relationship, the temporality of the events, and the biological gradient. Therefore, in order to draw appropriate conclusions, the survey design and the statistical modelling should be chosen carefully: longitudinal study should be preferred to cross-sectional surveys and multivariate models with random effects should be preferred to

▶

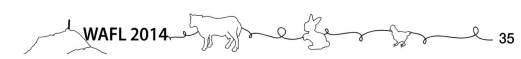

univariate models. Such methods can be used to study welfare disorders. Nevertheless, if a risk factor is often specific to a given disease, in animal welfare, a factor may influence different welfare aspects, sometimes differently. Basically, the knowledge produced with analytic surveys is mobilized to conceive action plans to be applied to a population of farms. In a farm, the suppression of the exposure to a risk factor will not necessary reduce the incidence of the disease. Indeed, a risk factor was evidenced at a population level, but in a given farm, the strength of the relationship between exposure to the risk factor and disease occurrence may be weak. Moreover, farmers have to comply with the action plan and to apply the corrective actions. For several reasons, the compliance with action plans to improve health is low, and even worse in the case of animal welfare. A type of analytic survey is particular relevant to assess the efficacy of a corrective action under field conditions, namely the controlled field trial. Like in experimental studies, the farms are randomly allocated to a 'treatment group', but the farms differ for all the other management practices, including the farmer. In such a longitudinal study design, the researcher should initially propose acceptable corrective actions and then, he can determine the factors affecting the compliance of the farmers and the efficacy of the corrective action. To conclude, operational and valuable knowledge can be produced on animal welfare using epidemiological concepts and methods. These methods do not allow studying animal welfare (or animal health) as a whole; rather they can help describe the status of farms regarding specific welfare disorders, and identify potential causal factors. This in turn is likely to help design corrective actions.

Main animal welfare problems in livestock during pre-slaughter operations: a Latin American view

C. Gallo

OIE Collaborating Centre for Animal Welfare and Livestock Production Systems Chile-Uruguay-Mexico; Animal Welfare Program, Faculty of Veterinary Science, Universidad Austral de Chile, Chile; cgallo@uach.cl

Animals for meat production are usually exposed to many stressors during production especially during pre-slaughter operations. Handling animals on farm, loading in /unloading from vehicles, transport, passing through livestock markets, fasting, lairage and stunning at the slaughterhouse can all affect welfare. How badly welfare can be affected will depend on intrinsic factors of the specific type of animals (species, sex, age, etc.) and on extrinsic factors of the environment where animals live or are handled, including the animal handlers. In Latin America, it has been strategic to address ethical aspects of improving animal welfare and to emphasize the close relationship between pre-slaughter handling of animals and the quantity and quality of meat produced. For large producers, considering animal welfare during production and pre-slaughter operations can determine the possibility of exporting or getting better prices for their products. At small farm level, particularly in less developed countries where human welfare is impaired, this strategy together with education has been relevant. It is important that education and training are done according to cultural, religious and socio-economical characteristics of each country. Therefore local research becomes relevant. In Chile and some other Latin American countries, some of the main problems encountered in cattle and sheep around slaughter are related to lack of proper infrastructure to handle animals; long distance transport with high stocking densities in the larger countries; long fasting times due to animals passing through livestock markets and dealers; bad handling of animals by untrained personnel in these and other premises; and finally lack of concern of operators in charge of stunning procedures. Positive results of local interventions at these stages have been used to train animal handlers and transporters, by showing them the consequences of bad handling with local audiovisual material. Research results have helped to support the development of new legislation or to make changes in the existent legislation related to animal welfare in several Latin American countries. During the last 15 years also consumer concern about the welfare of production animals in Latin America grew rapidly. Researchers in these countries have become increasingly involved in investigating and trying to solve animal welfare problems, leading to significant improvements in animal welfare accompanied by positive economic impacts that can lead to better human welfare.

Social attitudes and farm animal welfare

D.M. Weary and M.A.G. Von Keyserlingk
Animal Welfare Program, University of British Columbia, Vancouver, Canada; dan.weary@ubc.ca

Public interest in the welfare of farm animals is on the rise, but people working with these animals possess few mechanisms to discuss contentious issues with others in society. We review recent research using our Cow Views platform that creates virtual town hall meetings. This platform allows farmers, industry experts and the public to discuss controversial issues pertaining to dairy production, and provides a mechanism that enables different perspectives to surface regarding these issues. We discuss four issues: tail docking, pain control for disbudding, pasture access, and early separation of cow and calf. For each issue we used qualitative analysis to identify key themes raised by 100's of participants, including producers, veterinarians and people not working in agriculture. This analysis shows the diverse values that participants bring to the issue (e.g. some proponents of pasture access emphasize the importance of natural behaviour while some opponents emphasize the importance high levels of production), differences in beliefs about the 'facts' around an issue (e.g. some proponents of early cow-calf separation believe that the practice improves cow and calf health and some opponents believe the opposite), and perceived barriers to changing practice (e.g. some proponents of tail docking believe that there are no other effective means of maintaining good levels of cow hygiene on farm). Open discussion of contentious issues among farmers, industry professionals and the general public is an important step in the development of practices that better meet public expectations. The analysis of this discourse can also help identify priorities for research by animal welfare scientists, including science to address the competing factual claims and the development of management solutions that address the values of both farmers and the general public.

A split-sample experiment on citizen perceptions of intensive broiler fattening systems

G. Busch and A. Spiller
Georg-August-University of Göttingen, Department of Agricultural Economics and Rural Development, Platz der Göttinger Sieben 5, 37073 Göttingen, Germany; gbusch1@gwdg.de

Since the beginning of the 20th century former mixed family farms have developed into specialized production units. Broilers in particular are kept in big units and the intensive system showed more animal welfare problems compared to organic production. Animal welfare is also frequently discussed in the media and there is clear public concern over this issue in some European countries. These concerns consist of the worry about the well-being of the animals, but also negative impacts on food safety, quality and human health are feared from low animal welfare standards. Today, mass media are the main information source about agriculture for the general public. Mass media communication mainly relies on pictures, as they can be remembered faster and can transmit emotive messages better. In contrast, information from the agricultural sector is mainly provided via text. The present study investigates German citizens' perceptions of pictures from an intensive broiler fattening system (barn: 18×100 m; 40,000 animals) in order to gain information about how pictures can be used in communicating agricultural practice and welfare enhancements into public. A total of 291 citizens were asked via an online survey using a split-sample approach with five groups in July 2013. Each group was shown one of five pictures from the broiler barn. Four pictures showed different stages in the production cycle (day 1, day 7, day 34, day 40). A fifth picture was also taken at day 34 but varied in stocking density (43 kg/m² compared to 35 kg/m²). Via five point Likert-Scales, picture perception was tested. Data was analysed using IBM SPSS Statistics 21 (one-way analysis of variance (ANOVA) with post-hoc multiple mean comparison tests and Pearson bivariate correlations). The results show that the level of animal welfare parameters in the barn is evaluated to be low in general and no differences between stocking densities are perceived. Only for the split with the day-old chicks animal welfare was evaluated to be higher. Positive correlations between trust in the agricultural sector and better animal welfare evaluation in the shown barn can be observed, while meat consumption patterns do not influence picture perception. As a conclusion, new challenges for communication strategies arise from these results. It seems that the stocking densities considered appropriate in political and stakeholder discussions about animal welfare enhancements do not fulfil citizens' demands. Those campaigns are, although maybe better for the well-being of animals, not communicable. This holds also true for the common herd size of 40,000 animals.

Humans and farm animals at work: a multidisciplinary approach to study 'relational practices'

X. Boivin[1] and F. Kling-Eveillard[2]
[1]INRA, UMR1213 Herbivores, VetAgroSup, Centre de recherches, 63122, Saint-Genès-Champanelle, France, [2]Institut de l'Elevage, 149 rue de Bercy, 75012 Paris, France; xavier@clermont.inra.fr

Human-farm animal relationship is an important axis of the new European animal welfare policy. In our presentation we aim at showing the need of multidisciplinary approaches including social and biotechnical sciences to study a new concept of 'relational (human-animal) practices' on farms. It consists of practices that are voluntarily implemented by the farmer in order to improve humans-animals interactions, to get safety and efficiency for him(her)self, together with animal welfare. Following the ethological approach, the human-animal relationship is built from regular interactions between individuals that know each other. Animal responses to humans reflect this relationship. Several factors play a role such as genetic, environment, sensitive periods in the cycle of production, positive or negative impacts in the succession of the human-animal interactions.... Following psychological or sociological approaches, stockpeople attitudes, in particular behavioural beliefs about the efficiency of human behaviours towards the animals are now considered as key factors determining daily interactions. However, questions are still numerous, in particular concerning the way these interactions between humans and animals are taken into account in the work team organization. When something changes in the husbandry activity, for example automation, the concept of relational practices may help to understand what is at stake between human and animal, and how to cope with it in order to maintain or improve animal welfare and farmer working conditions. Crossing different disciplines, such as animal science, ethology, sociology of work and ergonomics, a systematic analysis of the human-animal interactions could be particularly useful on the design of work activity that suits the persons in that it takes account of their capabilities and limitations. Relational practices should be built on the basis of 'technical' choices depending of stockperson referential, (education, social transmission and social pressure), local and personal constraints (environment, resources, life choices) and available indicators of the way the husbandry system and the human animal relationship are running, including animals' response to humans. This multidisciplinary framework could make it possible to build an operational methodology to analyse the variability of farmers' relational practices and help them to achieve their goals on the farm. Some examples taken from cattle production will be presented.

Farmers' practices and attitudes to animal pain

F. Kling-Eveillard[1], V. Courboulay[2], M. Le Guénic[3], F. Hochereau[4], A. Selmi[5], M. Foussier[1] and A. Seigner[1]

[1]Institut de l'Elevage, 149 rue de Bercy, 75595 Paris cx 12, France, [2]IFIP-Institut du Porc, La Motte au Vicomte BP 35104, 35651 Le Rheu cedex, France, [3]Chambre d'Agriculture Bretagne, Avenue General Borgnis Debordes, 56002 Vannes, France, [4]INRA SenS, Université Paris Est, 5 bd Descartes, 77454 Marne-la-Vallée, France, [5]INRA UR 1326 SenS, Université Paris-Est, 5 bd Descartes, 77420 Champs sur Marne, France; florence.kling@idele.fr

A two-step qualitative survey was carried out to get a better knowledge of players' views and practices regarding pain management at farm level. Motivations and barriers of farmers to take pain into account were first explored through two focus groups (one in cattle and one in pig) comprising farmers' representatives, veterinarians, advisers and researchers. Then, 46 face-to-face semi-directive individual interviews were conducted with cattle and pig farmers. Farmers were asked about their monitoring practices with animals, their definition of animal pain, the situations causing pain on farm, their perception and practices related to pain during various routine management procedures, and their information needs about pain. The focus groups provided hypothesis for the farmers' interviews. The interviews were tape-recorded and a content analysis was performed to provide a description of the diversity of views. The results show that farmers are used to thinking about pain for their animals mainly within a context of disease, and they pay much attention to animal health. They often state that animal pain is complex to assess and they express a lack of knowledge concerning pain, its causes, its impact on animal and the signs given by a suffering animal (behaviour indicators for instance). According to cattle farmers, calves feel pain during the disbudding procedure, but it is brief and has no long-term consequences. According to pig farmers, piglet castration is painful whereas tail docking is diversely assessed, painful or not. Pain relief is performed by some farmers during the procedures, for different reasons: professional obligation, economic or technical advantage, profession image, animal welfare. The other farmers consider it to be excessive, ineffective, expensive or difficult to do (in terms of working duration, work organization, etc.). Three farmer profiles could be described in each species, related to their attitudes to animal pain and to their motivations or barriers to relieve it. It brings out that farmers are diversely sensitive to animal pain and prepared or not to relieve animal pain during disbudding, castration and tail docking. These profiles will be used as references for better targeting the training sessions and the good practice dissemination materials that are being elaborated as part of this 3-year project.

Conceptions of animal welfare and a 'good cow' in early modern rural Finland

T. Kaarlenkaski
University of Eastern Finland, School of Humanities, PL 111, 80101 Joensuu, Finland;
taija.kaarlenkaski@uef.fi

During the last thirty years, the discussion on animal welfare in both scientific and public discussion has increased remarkably. However, the history of the concept and early non-scientific perceptions of animal welfare have been underexplored. The purpose of this paper is to investigate the conceptualisations of animal welfare and good milking cows in the context of small-scale dairy farms in late 19th century and early 20th century Finland. The modernization of Finnish agriculture started in the 1860s as the focus of production shifted away from grain growing towards milk production, since the former had become unprofitable due to foreign trade and years of crop failure. Despite the gradual modernization process, traditional customs and ways of thinking persisted in many areas of Finnish society until the beginning of the 20th century. The period was characterized by the contradiction between traditional ideas and practices and, on the other hand, the information on new cattle tending methods distributed by the press, guide books and farming societies. The study is conducted by qualitative analysis using methods of textual research, including content analysis and close reading. The analysis focuses on different types of written materials produced between the 1860s and 1930s. These include answers to an ethnographic questionnaire about cattle husbandry, cattle tending guide books and articles in newspapers. The materials show that the concept of animal welfare was referred to in newspaper articles related to cattle tending as early as in the 1890s. Guidance on how to improve animal welfare was given in cattle tending guide books published in 1907 and 1923. According to these books, the most important sectors of animal welfare were light, fresh air, warmth, cleanliness of both the animals and the cowshed, proper treatment of the cloven hoofs, orderliness in cattle tending and good treatment of the animals. In vernacular thought the conceptions of good milking cows were somewhat different from present-day ideas. Abundant milk production was not the only desirable quality of the cow; equally important was that it did not eat too much and came home easily from the pastures in the forest. In addition, traditional beliefs suggested that also the appearance of the animal had to be taken into account when choosing the calves to be raised as milking cows. In conclusion, the materials show that there was interest in the questions of animal welfare and the qualities of farm animals already more than one hundred years ago. These issues were an integral part of everyday work on small-scale farms and were discussed and redefined along with the process of modernization in cattle husbandry.

Involving stakeholders in tail biting prevention: stakeholder opinion on a German tail biting tool

D. Madey[1], A.L. Vom Brocke[1], M. Wendt[2], L. Schrader[1] and S. Dippel[1]
[1]Friedrich-Loeffler-Institut, Institute of Animal Welfare and Animal Husbandry, Doernbergstraße 25/27, D-29223 Celle, Germany, [2]University of Veterinary Medicine, Foundation, Clinic for Swine and Small Ruminants, Bischofsholer Damm 15, 30173 Hannover, Germany; dana.madey@fli.bund.de

Tail biting is a behavioural disorder in pigs which reduces animal welfare and productivity. Involving farmers and advisers in its prevention poses a key factor for successful tail biting reduction. This contribution presents stakeholder opinion on the tail biting management and advisory tool SchwIP, which is currently being evaluated in a research project in Germany. SchwIP combines risk assessment with herd health and welfare planning. The automatically generated farm specific report with background information and suggestions for improvement is based on a database with knowledge from science and practical farming. The first author as well as advisers and veterinarians trained on the tool (AV) applied SchwIP twice at an interval of one year on 188 German fattening farms. Afterwards, 48 of the AV and 146 of the visited farmers answered an anonymous opinion questionnaire. Almost all AV and farmers regarded SchwIP as ready for practical application (92% of AV, 98% of farmers). Seventy-nine percent of AV stated that they would be able to integrate SchwIP in their daily work. The measures suggested by the tool were regarded as realistic by 98% and useful by 95% of farmers. Seventy-one percent of farmers and 55% of AV learned more about tail biting causation by using the tool, and 79% of farmers and 69% of AV gained a new overview of tail biting causes on their respectively the customer farm. Knowledge transfer is the most plausible explanation for the relatively high implementation rate of 79% of farmers carrying out at least one measure in the study period, as 96% of these farmers implemented measures because they had been convinced of their effectiveness. Management and advisory tools combining risk analysis, background knowledge and recommendation of tail biting prevention measures can therefore help in raising awareness of the issue in farmers and advisers. This in turn encourages implementation of risk reducing changes.

A Delphi application to define acceptability levels for welfare measures during long journeys

H. Spoolder[1], V. Hindle[1], P. Chevillon[2], M. Marahrens[3], S. Messori[4], B. Mounaix[5], C. Pedernera[6] and E. Sossidou[7]
[1]WUR Livestock Research, Dept Animal Welfare, P.O. Box 65, 8200 AB, Lelystad, the Netherlands, [2]IFP, 3-5 rue Lespagnol, 75020 Paris, France, [3]FLI, Doernbergstr. 25-27, D-29223 Celle, Germany, [4]IZSAM, Campo Boario, 64100 Teramo, Italy, [5]Institut de l'Elevage, 35652 Le Rheu, Cedex, France, [6]IRTA, 17121, Monells, Spain, [7]Hellenic Agricultural Organization-DEMETER, 57001, Thermi-Thessaloniki, Greece; hans.spoolder@wur.nl

Future welfare certification systems concerning long journeys may incorporate the use of animal based measures (ABMs). During the DG SANCO funded European project 'Quality Control Posts', sets of appropriate ABMs were identified for welfare assessment for transport of sheep, cattle, horses and pigs. In the present study, we aimed to establish expert consensus on acceptability thresholds for ABMs, in order provide an aggregate overall score, using a two phase Delphi procedure. In Phase 1, experts were asked to identify two threshold (TH) levels per measure: values <TH1 were considered 'certified', and values >TH2 were 'unacceptable'. Values between TH1 and TH2 were 'acceptable' but not preferred in a certification scheme. Phase 1 consisted of 3 rounds. In round 1 experts were asked to provide their opinion on TH's, in round 2 they responded anonymously to opinions of other experts. In round 3 a final decision based on the anonymised arguments was comprised. During phase 2 all experts were approached again for their maximum numbers of 'acceptable' measures, in order for a species set of measures to maintain 'certifiable' status. This was accomplished in 2 rounds. Respectively, 185, 106 and 91 experts representing 10 different countries provided their opinions on thresholds in the 3 rounds of phase 1. Results suggest that variation in threshold values decreases in subsequent rounds: e.g. a reduction in difference between 1st and 3rd quartile over three successive rounds occurred in 18 out of 20 TH values for the set of 10 sheep ABMs (2/20 ABMs; Sign Test, P<0.001). Phase 2 provided 127 and 101 respondents in round 1 and 2 respectively. The maximum 'acceptable' ABMs for a set to be approved as 'certifiable' differed slightly between species: 3/11 weaner pigs, 4/12 finishing pigs, 3/13 calves, 4/14 heifers, 3/14 other cattle, 4/14 horses, 3/10 lambs and 4/10 adult sheep. Across all species, the variation in replies decreased, as the difference between 1st and 3rd quartile was lower in round 2 compared to round 1 (8/8 sets; Sign Test, P<0.001). In conclusion, the Delphi procedure helped to achieve expert consensus and the resulting threshold values are recommended for use in the development of a welfare certification system.

A pain scoring protocol for dairy cattle

K.B. Gleerup[1], P.H. Andersen[2], L. Munksgaard[3] and B. Forkman[1]
[1]University of Copenhagen, Department of large animal sciences, Gronnegaardsvej 8, 1870 Frederiksberg, Denmark, [2]Swedish University of Agricultural Sciences, Department of Clinical Sciences, Ulls väg 12, 750 07 Uppsala, Sweden, [3]University of Aarhus, Department of Animal Science, Blichers Allé 20, 8830 Tjele, Denmark; kbg@sund.ku.dk

In many animal welfare protocols, pain is assessed through measuring conditions that are thought to be related to pain, e.g. wounds or lameness. New developments in the field of pain assessment has shown that it is possible to measure acute pain in a more direct manner via the facial expressions and behaviour of the animal. In the current study we developed a pain scoring protocol for dairy cattle, including observations of attention, head bearing, ear position, facial expression, back position and response to approach. The behaviour of 43 dairy cows was assessed and based on a clinical examination they were divided in a pain group of 23 cows and a control group of 20 cows. All cows were treated randomly with either an analgesic (ketoprofen) or a placebo (saline). After 2-4 hours, the animals were re-assessed according to the pain scoring protocol by an observer blinded to the treatment of the animals. There was a significant difference before treatment, in the overall pain score between the pain and the control group (Mann-Whitney $P=0.0001$) and for the pain group, there was a significant reduction in the pain score of the ketoprofen but not the saline group (Wilcoxon $P=0.006$; $P=0.05$). In conclusion, the current pain protocol based on simple short term observations of behaviour has potential for assessment of pain in dairy cattle on farm.

Monitoring in cattle and horse for the determination of animal health, animal welfare and well-being

H.-U. Balzer, K. Kultus and S. Köhler
Humboldt Universität zu Berlin, Institut für Agrar und Stadtökologische Projekte, Philipp Str. 13, 10115 Berlin, Germany; ullrich.balzer@agrar.hu-berlin.de

The aim of the project was the development of criteria for a complex objective assessment of animal welfare from housing systems on the basis of individual animal information to be detected during the lifetime of the individual. The assessment is namely based on the health state of the animals, covering all husbandry influences. For this purpose, an innovative method should be developed, allowing the determination of the health status by a rating scale on the basis of the monitoring of psycho-physiological parameters and behavioural parameters. Concerning dairy cattle, three different housing systems were studied at 8 livestock areas (cubicle housing, deep litter and tethering). A total of 109 cattle was investigated at 1024 study days, with an average duration of measurement time of 9.4 days/animal. In the case of horses, three different housing systems were studied at 11 livestock areas. (group housing, housing in groups of two animals, single housing). 90 horses were investigated at 811 study days, with an average duration of measurement time of 9.0 days/animal. For the assessment of animal welfare and animal health, the following evaluation methods were used. (1) National Framework for Evaluation of Animal Husbandry Procedures (NBR); (2) Animal Welfare Index for cattle TGI 35 L/1996; (3) Veterinary Medical Status; (4) Investigator´s assessment; (5) Animal status by use of animal monitoring; (6) Behaviour assessment through video recordings. The determination of the animal status was performed by monitoring of vegetative parameters: vegetative-emotional response, vegetative-nervous response, muscle activity, information on the metabolic reaction and the behavioural parameters by means of the smardwatch®- measurement system. Data analysis was performed using the Chronobiological Regulation Diagnostics (CRD). As result of the project, a new method for determining the animal health was developed as a future basis for the objective assessment of animal welfare by means of monitoring vegetative parameters. For the first time main characteristics – vegetative and regulatory balance – were calculated by use of the CRD-method. The comparison of the above-mentioned new assessment method for animal welfare with the animal status in cattle showed an accordance of about 60% with the veterinary medical status and contradictory results with those of the other valuation methods (NBR, TGI, Welfare Quality®). In horses, similar differences were observed. In addition, investigations were carried out for stress type, the sensibility and the human – animal behaviour.

Heart rate variability: a biomarker of dairy calf welfare

J. Clapp
Newcastle University, School of Agriculture, Food & Rural Development, Agriculture Building
Newcastle University Newcastle, NE1 7RU, United Kingdom; james.clapp2@ncl.ac.uk

Dairy calf welfare is recognised to be compromised from common management practices. In this study heart rate variability (HRV) was used to measure stress in 25 young dairy calves to quantify the degraded welfare they experienced from weaning separation and isolation also the painful disbudding procedure. It was shown the time spent on the cow before separation had a significant negative correlation to HRV (r^2=-0.68, P=0.03). The longer a calf spent in isolation the lower its HRV three days after joining a group pen (P=0.037). The removal of a dummy teat elicited a significant drop in HRV (P=0.05), identifying the addictive properties of sucking in claves. Post disbudding stress, reflected by declining HRV values, was only partly alleviated by the NSAID meloxicam after 48 h. The findings showed calf welfare would be improved by reducing the time between birth and separation also the days spent in single pens. Providing dummy teats for individually housed calves showed potential as a positive environmental enrichment. Meloxicam may improve welfare by alleviating some chronic pain following hot iron disbudding. We conclude these findings illustrate that HRV, as a science-based animal-centric biomarker of animal welfare, should be more widely used to improve farmed animal practice.

Characteristics of eye surface temperature in dairy cows measured by infrared thermography

P. Savary[1], S. Kury[1,2] and T. Richter[2]
[1]Agroscope, Tänikon 1, 8356 Ettenhausen, Switzerland, [2]University of Nürtingen-Geislingen, Neckarsteige 6-10, 72622 Nürtingen, Germany; pascal.savary@agroscope.admin.ch

Most stress analysis methods require an invasive procedure that can affect the animal's physiological and behavioural state. Studies have shown that a drop in eye temperature, measured using an infrared camera, can be an indicator of acute stress in an animal. Thermography is a non-invasive method, but temperature variations can be caused by other factors besides stress. This study aimed to quantify the effects of the position of the camera and the eye surface temperature fluctuation during a short period in non-stressed dairy cows. The subjects of the present study were twelve lactating cows. During feeding (9:00 am) eye surface temperature was measured frontally and laterally with a distance of 0.5 m, 1.0 m and 1.5 m between the camera and the eye. Maintaining a distance of 0.5 m the temperature was measured three times in the space of one and a half minutes, during milking (4:30 am and 4:00 pm) and when the cows were lying down (6:00 am, 12:30 pm and 6:30 pm). Measurements were made on two successive days. The data were analysed statistically using generalised linear mixed models. The surface temperature decreased significantly ($P<0.001$) with increasing distance between the camera and the eye. With frontal recording the temperature values were 0.75 °C lower than with lateral ($P<0.001$). The eye temperature did not vary significantly during the ninety-second period ($P=0.594$). However a circadian rhythm of the eye surface temperature was observed when cows were lying down. The surface temperature increase significantly ($P<0.001$) from the morning to the evening measurements. These results show that, based on the constant eye surface temperature during a short period and taking into account a constant distance and angle between the camera and the eye, this method can be used to analyse acute stress in dairy cows.

Inter-relations between animal-based measures of welfare of dairy cows and effects of housing type

A.M. De Passille[1], J. Rushen[1], E. Vasseur[2], J. Gibbons[3], D. Pellerin[4], V. Bouffard[4,5], D.B. Haley[2], J. Heyerhoff-Zaffino[2], C. Nash[2], D. Kelton[2] and S. Leblanc[2]
[1]University of British Columbia, Agassiz, BC V0M 1A0, Canada, [2]University of Guelph, Guelph, ON N1G 2W1, Canada, [3]Dairy Co. UK, Kenilworth, Warwickshire CV8 2TL, United Kingdom, [4]Laval University, Quebec City, QC G1V 0A6, Canada, [5]Valacta Inc., Ste-Anne-de-Bellevue, QC H9X 3R4, Canada; passille@mail.ubc.ca

Increased use of animal-based welfare standards is partly due to the fact that they allow comparison between different types of housing systems that would be difficult using input-based measures. However, taking animal-based measures is time consuming and we need to identify any redundancy in these measures to limit their numbers. To examine relationships between various animal-based measures of dairy cow welfare and the differences between housing systems, we measured the prevalence of lesions to the hocks, knees and the neck, lameness, body condition scores (BCS), and dirtiness on 40 lactating Holstein cows from each of 80 free stall farms (FS) and 100 tie-stall farms in two Canadian provinces, Quebec (QC) and Ontario (ON). The prevalence of all measures varied greatly among farms. The prevalence of hock injuries on a farm was correlated (P<0.001) with the prevalence of knee lesions (r=0.56), neck lesions (r=0.36) and lameness (r=0.39). The prevalence of dirty cows and cows with low BCS were not correlated with other measures. There were no associations with farm size or farm annual milk yield, except in free-stalls where the prevalence of low BCS was negatively correlated with farm annual milk yield (r=-0.38; P<0.01). Differences between housing types were complicated by interactions between housing type and province. Neck injuries were more prevalent (P<0.05) in tie-stalls (mean ± SE=40.0±2.1% cows) than free-stalls (11.5±3.0% cows) in both provinces. Dirty cows were more common (P<0.05) in ON (21.7±1.8% cows) than QC (11.1±2.1% cows) in both housing systems. Lameness was more common in tie-stalls (33.5±1.5% cows) than free-stalls (24.3±2.5% cows) in QC (P<0.05), but was higher in free-stalls (21.9±1.8% cows) than tie-stalls (10.9±1.8% cows) in ON (P<0.05). Large differences between farms in all measures were far greater than differences between farm types and differences between provinces suggest a possible effect of local advisory or extension services. Hock injuries are easily measurable and a high prevalence on a farm may indicate other injuries and lameness, and possibly be used in an initial screening of farms.

Investigating integument lesions in dairy cows: which types and locations can be combined?

C. Brenninkmeyer[1], S. Dippel[2,3], J. Brinkmann[4], S. March[4], C. Winckler[2] and U. Knierim[1]
[1]University of Kassel, Farm Animal Behaviour and Husbandry Section, Nordbahnhofstraße 1a, 37213 Witzenhausen, Germany, [2]University of Natural Resources and Life Sciences, Division of Livestock Sciences, Gregor-Mendel Straße 33, 1180 Vienna, Austria, [3]Friedrich-Loeffler-Institute, Institute for Animal Welfare and Animal Husbandry, Dörnbergstraße 25 + 27, 29223 Celle, Germany, [4]Federal Research Institute for Rural Areas, Forestry and Fisheries, Thuenen-Institute of Organic Farming, Trenthorst 32, 23847 Westerau, Germany; brennink@uni-kassel.de

In studies of integument lesions (IL) in dairy cows, different types of lesions (e.g. wounds and swellings) and different locations on the cows' body (e.g. all legs or the whole hock area) are frequently analysed in one statistical model, assuming that they have common underlying causes. In studies differentiating between types and locations (e.g. the tuber calcis from the tarsal joint), it became apparent that different risk factors can be identified. To systematically investigate which types and locations correlate and may thus be combined, a hierarchical cluster analysis on herd level prevalence of three different types of IL (hairless patches, scabs and wounds, swellings) at 13 frequently affected body parts was performed. The sample comprised 95 conventional and organic dairy farms, with data collected in the winter housing period 2004/05 in Germany and Austria from 2,922 lactating Holstein Friesian and Simmental cows. A high increase in explained variance was achieved at a 2, 5 and 14 cluster stage, explaining 38, 55 and 81% of the inter-herd variance, respectively. The 2 cluster stage roughly divided the lower legs from the upper body parts. Further clustering separated the medial hock area from the rest of the lower legs. For the upper part of the body, one cluster represented front parts, one rear parts and one a mixture of front, rear and middle. The 14 cluster stage in some cases revealed differentiation between swellings and the other IL types and combined locations at the tuber calcis with the carpal joints, separating it from the lateral tarsal joint. Results show that location overrides type of IL in most cases and that sometimes, clusters represent widely spaced body parts, e.g. the carpal joint and lateral and dorsal tuber calcis, whereas neighbouring areas as the medial and lateral hock area should be analysed separately from each other for causal analysis.

A study into the performance of *in vivo* methods to assess osteochondrosis in growing piglets

C.P. Bertholle
Utrecht University, Farm Animal Health, Yalelaan 7, 3584 CL Utrecht, the Netherlands; c.p.bertholle@uu.nl

Leg weakness is a significant source of both welfare and economical issues in today's porcine industry, and is known to be mostly caused by osteochondrosis (OC), a disease affecting cartilage during growth. The etiological factors of this disease have been studied in depth, however little is known on its effect on locomotion and gait and whether these effects can be detected by methods that have not yet been validated for pigs. The study focuses on the detection of early lesion processes linked to OC. This will be attempted by using in-vivo methods like visual scoring of locomotion and gait, pressure plate measurements and X-rays. The end goal is to enable professionals in the porcine industry to be able to diagnose the disease accurately, non-invasively and in the highest through-put way possible. For this study, we examined nineteen weaned piglets longitudinally for the presence of OC. Histology was used as the gold standard. The in-vivo techniques tested included a pressure plate analysis technique, designed to measure a range of kinetic factors, and also a visual scoring method already designed and tested for pigs. The experiment was divided into four time-points all separated by an interval of three weeks. At the first three time-points, three pigs were euthanized and examined pathologically as well as histologically for OC lesions at specific predilection sites (elbow, stifle and hock). At the final time-point the remaining ten pigs were similarly euthanized and examined. All pigs, while alive, were examined weekly with the visual scoring method and pressure plate measurements. Only the final remaining ten pigs were followed with X-rays. The reproducibility of the in-vivo techniques was assessed using inter rater reliability tests. The ability of the in-vivo techniques to detect OC in these early age pigs was evaluated using a Principal Component analysis (PCA). Both in-vivo techniques, visual scoring and pressure plate analysis, were reproducible and consistent in between measuring sessions. Visual scoring demonstrated a 60% rate of agreement, between the raters, for three quarters of the visual scoring factors scored. The results from the pressure plate analysis showed an equally encouraging result in reproducibility with ICC 'Two way consistency' results ranging from 0.463 to 0.595. Finally, both in-vivo techniques had their results categorized into positive and negative OC groups depending on their corresponding X-ray results, and were evaluated in the PCA statistical analysis. The results from the PCA suggested that results in the in-vivo techniques are able to be dissociated into OC negative and OO Positive ranges, confirmed by X-ray analysis.

Can tear staining in pigs be useful as an on-farm indicator of stress?

H. Telkänranta[1], A. Valros[1] and J.N. Marchant-Forde[2]
[1]Research Centre for Animal Welfare, P.O. Box 57, 00014 University of Helsinki, Finland, [2]USDA-ARS, LBRU, 125 S Russell St, West Lafayette, IN 47907, USA; anna.valros@helsinki.fi

Tear staining (TS) is a reddish-brown eye secretion, used as a stress measure in rodents. TS in pigs is often considered a result of poor air quality or respiratory disease. An increase in TS has also been observed in pigs exposed to stressful situations, and has been connected to physiological stress measures. The current aim was to evaluate TS as an on-farm measure of stress. A 0-5 scale (no staining to extensive staining) was used to quantify TS during live observations as part of a larger study comparing effects of manipulable objects on pig welfare. Three commercial farms were included, one fattening (Farm 1), one growing breeding gilts (Farm 2) and one integrated, including both the farrowing and fattening units (Farm 3). Observations on tail- and ear damage (fattening pigs, gilts) were performed, as well as a novel object test (piglets) and a novel person test (gilts). Pearson or Spearman correlations were used to test the interaction between stress-related measures and TS and independent t-tests or mixed models with pen as random factor to test for treatment-effects (SPSS 21). Treatment (control, plastic object, wood object) tended to affect left eye, but not right eye TS on fattening pigs on Farm 3 (F_{41}=2.8, P=0.07), with less TS in pigs from pens with plastic or wooden manipulable objects than in control pens (2.7, 2.8 and 3.0 resp.). On Farm 1 there was no overall effect of treatment (control, branched chain, plastic object, wood object, combination of all), but pairwise comparisons showed less TS in pigs from pens with the combination of all objects as compared to control pens (3.3 vs 3.9, F_{29}=4.2, P<0.05). On Farm 3, sows, whose litters had been given additional sisal ropes, tended to have less TS than control sows (t=-1.9, P=0.07) and TS correlated with latency of piglets to approach a novel object in their home pen (29 litters, rp=0.41, P=0.03). On Farm 1, average TS of both eyes (n=780) correlated with tail damage score (rs=0.14, P<0.001) and ear damage score (rs=0.16, P<0.001). Also on Farm 3, where TS was scored separately for each eye, TS correlated with tail damage score (n=654, left eye: rs=0.12, P=0.002; right eye: rs=0.07, P=0.06), while ear damage score correlated with right eye TS only (rs=0.1, P=0.009). On Farm 2 TS tended to correlate with pen-level latency to approach an unfamiliar person (n=44, rp=0.27, P=0.07 for both eyes). TS was affected by treatment, indicating that it is related to long-term welfare. For most measures, left eye TS yielded more results. The potential of TS as an on-farm indicator of pig welfare is further supported by the correlation to several other stress-related measures.

Monitoring procedures at slaughterhouses for pigs

A. Velarde[1], C. Berg[2], M. Raj[3], H.H. Thulke[4], D. Candiani[5], M. Ferrara[5], C. Fabris[6] and H. Spoolder[7]
[1]IRTA, Veïnat de Sies s/n, 17111, Monells, Spain, [2]SLU, P.O. Box 234, 53223 Skara, Sweden, [3]University of Bristol, Langford, BS40 5DU, United Kingdom, [4]Helmholtz Cenntre for Environmental Research, Permoserstr 15, 04318, Leipzig, Germany, [5]EFSA, Via Carlo Magno 1/A, 43121, Italy, [6]Regione del Veneto, Dorsoduro 3493, 30125 Venezia, Italy, [7]WUR Livestock Research, P.O. Box 65, 8200 AB Lelystad, the Netherlands; antonio.velarde@irta.cat

This study by the EFSA AHAW panel set out to develop toolboxes of welfare indicators for developing monitoring procedures at slaughterhouses for pigs stunned with head-only electrical method or carbon dioxide at high concentration. In particular, it proposes welfare indicators together with their corresponding outcomes of consciousness, unconsciousness or death. The toolbox of indicators is proposed to be used to assess consciousness at three key stages of monitoring: (1) after stunning and during shackling and hoisting; (2) during sticking; and (3) during bleeding. Various activities, including a systematic literature review, an online survey using a questionnaire, stakeholders' and hearing experts' meetings, were conducted to gather information about specificity, sensitivity and feasibility of the indicators that are to be included in the toolboxes for monitoring welfare. On the basis of information gathered during these activities, a methodology was developed to select the most appropriate indicators that could be used in the monitoring procedures. The frequency of checking differs between the roles of different people with responsibility for ensuring animal welfare at slaughter. The personnel performing stunning, shackling, hoisting and/or bleeding, will have to check all the animals and confirm that they are not conscious following stunning. For the Animal Welfare Officer, who has the overall responsibility for animal welfare, a mathematical model for the sampling protocols is proposed, giving some allowance to set the sample size of animals that he/she needs to check at a given throughput rate (total number of animals slaughtered in the slaughterhouses) and tolerance level (amount of potential failures; animals that are conscious after stunning; animals that are not unconscious or not dead after slaughter without stunning). The model can also be applied to estimate threshold failure rate at a chosen throughput rate and sample size. Finally, different risk factors and scenarios are proposed to define a 'normal' or a 'reinforced' monitoring protocol, according to the needs of the slaughterhouse.

Horse welfare assessment after long journey transports

S. Messori[1], M. Buonanno[1], P. Ferrari[2], S. Barnard[1], M. Borciani[2], N. Ferri[1] and K. Visser[3]
[1]Istituto Zooprofilattico Sperimentale dell'Abruzzo e del Molise G. Caporale, Campo Boario, 64100 Teramo, Italy, [2]Centro Ricerche Produzioni Animali – CRPA, Viale Timavo, 42121 Reggio Emilia, Italy, [3]Wageningen UR, Livestock Research, P.O. Box 65, 8200 AB Lelystad, the Netherlands; s.messori@izs.it

Horse meat consumption is considerable in Italy. The insufficient national breeding production and the consumer demand for live animals as opposed to carcasses engender a relevant trade of horses, often travelling over long journeys (more than 8 h). The transport of live animals, causing emotional and physical stress, gives rise to public concern about their welfare, particularly when transportation involves long journeys. Aim of this study was to develop a feasible post-transport horse welfare assessment protocol and to test it during commercial transports in Italy, to provide an insight of travel condition for these animals. A set of animal-based (ABM), resource-based (RBM) and management-based (MBM) measures, was developed, covering all welfare aspects of both transport (TR) and unloading procedure (UN). The protocol was tested on 51 commercial horse transports arriving to different sites in Italy. Univariate analysis was carried to look for associations between the outcome variable (ABM) and the input variables (RBM and MBM). Non parametric tests were applied according to the type of data, i.e. Spearman correlations for numerical continuous data, Wilcoxon test and Kruskall-Wallis test for categorical data. No severe welfare impairments were recorded (i.e. dead on arrival, severe injuries, non-ambulatory animals). For TR, no correlation emerged between the outcome variables and input variables ($P>0.05$), also due to the prevalence proximate to zero of most ABMs. At UN, inappropriate handling was positively correlated with reluctance to move (hitting animals: $P=0.002$ and excited movements: $P=0.03$) and fast gait ($P=0.05$ when handlers made loud noises). Ramp covering was associated to slipping events (animal slipped less when the ramp was covered with rubber mat compared to corrugated anti-slip metal, $P=0.002$), while ramp slope was positively correlated to falling ($P=0.005$) and fast gait ($P=0.03$). Longest travels appeared to be correlated with an increase in falling ($P=0.064$) and reluctance to move ($P=0.069$) percentages. The general health condition of horses arriving in Italy after long journey travels appeared to be sufficient. The protocol was found feasible and sensitive in identifying criticalities especially during unloading procedures. The protocol is ready to use on a larger sample, with more variety in season, distance travelled and origin and would be valuable in better monitoring the welfare of horses during transport and unloading.

Can automated measures of lying time help detect lameness and leg lesions on dairy farms?

G. Charlton[1], V. Bouffard[2], J. Rushen[3], J. Gibbons[1], E. Vasseur[4], D.B. Haley[4], D. Pellerin[5] and A.M. De Passille[3]

[1]Agriculture and Agri-Food Canada, Agassiz, BC V0M 1A0, Canada, [2]Valacta Inc., Ste-Anne-de-Bellevue, QC G1V 0A6, Canada, [3]University of British Columbia, Agassiz, BC V0M 1A0, Canada, [4]University of Guelph, Guelph, ON N1G 2W1, Canada, [5]Laval University, Quebec City, QC H9X 3R4, Canada; rushenj@mail.ubc.ca

The time that dairy cows spend lying down is an important measure of their comfort, and lameness and injuries to hocks and knees are associated with alterations in lying time. Automated methods now allow for the efficient measurement of lying time of large numbers of cows on commercial farms. We examined whether automated measures of lying time could identify cows and farms with lameness or leg lesion problems. Data was collected from 40 lactating Holstein cows from each of 100 tie-stall dairy farms. The occurrence of lameness, and hock and knee injuries was recorded, and lying times were automatically recorded over 3 d using accelerometers. There was large variation between individual cows, and between farms in all measures of lying time. There was no relationship (Logistic regression: P>0.10) between a cow being lame and the daily duration of lying time. A lower daily duration of lying time was found among cows with hock injuries (mean ± SE: non-injured=12.79±0.06 h, injured=12.21±0.06 h; Logistic regression: P<0.001) and knee injuries (non-injured=12.54±0.05 h, injured=12.25±0.06 h; Logistic regression: P=0.04). The median daily duration of lying time on a farm was negatively correlated with the prevalence of lameness (r_p=-0.27, P=0.006), of hock injuries (r_p=-0.35, P<0.001) and of knee injuries (r_p=-0.28, P=0.005). Farms that had a median daily duration of lying time of more than 12.5 h had a reduced odds of being above the median for hock lesion prevalence (Logistic regression: OR=0.37, P=0.02), a reduced odds of being above the median for knee lesion prevalence (Logistic regression: OR=0.29, P=0.003), and tended to have a reduced odds of being above the median for lameness prevalence (Logistic regression: OR=0.47, P=0.07). A criterion of a median lying time of 12.5 h or more could identify 69.0% of farms above and 66.7% of the farms below the median for lameness, hock, and knee lesion prevalence. Automated measures of lying time may be a useful animal-based measure to identify farms with a high percentage of lame or injured cows, and can increase our ability to collect behavioural measures in on-farm animal welfare assessments.

Behavioural changes after mixing can be detected in automatically obtained activity patterns of pigs

S. Ott[1,2], C.P.H. Moons[2], M. Kashiha[3], B. Ampe[4], J. Vandermeulen[3], C. Bahr[3], D. Berckmans[3], T.A. Niewold[1] and F.A.M. Tuyttens[2,4]
[1]*KU Leuven, Livestock-Nutrition-Quality, Kasteelpark Arenberg 30, 3001 Heverlee, Belgium,* [2]*Ghent University, Animal Nutrition, Genetics and Ethology, Heidestraat 19, 9820 Merelbeke, Belgium,* [3]*KU Leuven, M3-Biores, Kasteelpark Arenberg 30, 3001 Heverlee, Belgium,* [4]*Institute for Agricultural and Fisheries Research (ILVO), Animal Sciences Unit, Scheldeweg 68, 9090 Melle, Belgium; sanne.ott@biw.kuleuven.be*

Automated image analysis to measure animal behaviour facilitates the collection of daily and continuous data of animals. Continuous monitoring could be useful in livestock husbandry because certain changes in the baseline of daily activity patterns could indicate welfare and health problems. This study investigated the possibility to automatically detect changes in the activity patterns of pigs after mixing. Forty grower pigs (25.1±4.4 kg) were equally distributed over four conventional pens and balanced for sex, live weight and litter. After seven days, five pigs from one pen were swapped (07:00 h) with five from another pen (treatment pens T1 and T2). The two remaining pens served as controls (C1 and C2). Video images from a top view camera were collected and analysed for 10 hours (08:00 to 18:00 h) on four days before and four days after the treatment pens were mixed. Automated activity patterns of the control and treatment pens were obtained by calculating the mean activity index per minute (AI): the relative number of moving pixels between two consecutive image frames (1 frame/second) averaged per minute. Changes in the activity pattern of each pen were expressed as the sum of sudden deviations (SSDEV) per day. A sudden deviation was defined as a significant increase or decrease (significance level of 0.05) in the AI compared to the mean AI from the preceding 3 minutes. Within pens the SSDEV was comparable during the four recording days prior to mixing, but the SSDEV differed between pens: C1 43±5.0 (mean±sd); C2 30±2.5; T1 49±5.8; T2 28±2.2. On the day of mixing, the treatment pens both showed a two-fold increase in their SSDEV (T1 105; T2 77) compared to the days before and after mixing, while the control pens did not (C1: 36; C2:24). This likely reflected the behavioural changes (aggressive interactions) caused by the mixing. These preliminary results suggest that automated measurements of daily activity in pigs can be useful in detection of behavioural changes at pen level.

Pig slaughter: no signs of life before dressing and scalding?

S. Parotat[1], S. Arnold[2], K. Von Holleben[1] and E. Luecker[3]
*[1]bsi Schwarzenbek, Postbox 1469, 21487 Schwarzenbek, Germany, [2]Federal Research Institute
of Nutrition and Food, MRI Kulmbach, E.-C.-Baumann-Straße 20, 95326 Kulmbach, Germany,
[3]University of Leipzig, Faculty of Veterinary Medicine, Institute of Food Hygiene, An den
Tierkliniken 1, 04301 Leipzig, Germany; sp@bsi-schwarzenbek.de*

Previous studies have shown that stunning and sticking do not always induce irreversible unconsciousness. While EU regulation 1099/2009 requires business operators to verify the absence of signs of life before further processing no method for automated individual testing has been established so far. Therefore this study aims to lay groundwork for the development of an automated method to verify that death has occurred. The hypothesis is that following stunning and bleeding reactions to a painful stimulus can be triggered in some pigs signifying an at least partly functional brain. The stimulus as well as the detection of triggered movements could be automated preventing these pigs to be processed further without anew stunning. A total of 24,390 slaughter pigs from three commercial German abattoirs (A-C) were examined on line for reactions to an automated hot water stimulus applied on average 148 (92-348) seconds post sticking. Plants used modern CO_2-stunning systems operating at 420 to 520 pigs/hour. Sample sizes were 5,627(A), 12,302(B) and 6,461(C) respectively. Hot water (60 °C) was sprayed for four seconds onto the muzzle, head and upper body of each pig using a purpose built device. Behaviour during water exposure was observed. Pigs were then manually examined for the presence of reflexes (dazzle, corneal/palpebral) and reaction to a nasal septum pinch. Pigs with reactions/reflexes were captive bolt stunned. All pigs were videotaped between sticking and further processing. A share of 0.16% showed movements during hot water exposure (0.00%(A), 0.32%(B), 0.02%(C)). We observed blinking (87.5%), mouth opening (80.0%), grimacing (62.5%), head shaking (42.5%), righting reflex (45.0%) and foreleg movements (27.5%). Vocalization occurred in 7.5% of the cases. Reflexes and/or reactions to the nose pinch were present in 80.0% of pigs with movements (dazzle: 60.0%, corneal/palpebral: 77.5%, nose pinch: 37.5%). In 35.0% all three tests were positive. Pigs without movements in hot water also had negative results during manual testing. Current slaughter regimes involving CO_2-stunning carry a risk of sensible and/or conscious animals being further processed. While obvious differences were observed between abattoirs the overall incidence was low, emphasizing the need for automated individual detection. The results are discussed with regard to welfare relevance, quality of the slaughter process and possibility of developing automated vision systems.

Understanding feeding behaviour and associated welfare of pigs from a motivational perspective

I.J.M.M. Boumans[1], E.A.M. Bokkers[1], G.J. Hofstede[2] and I.J.M. De Boer[1]
[1]Wageningen University, Animal Production Systems Group, P.O. Box 338, 6700 AH Wageningen, the Netherlands, [2]Wageningen University, Information Technology Group, P.O. Box 8130, 6700 EW Wageningen, the Netherlands; iris.boumans@wur.nl

Social competition for feed among pigs or inadequate feed quality, can reduce health, increase stress and may result in abnormal behaviours such as tail biting or stereotypies and therefore impair pig welfare. Gaining more insight in behavioural responses of pigs to obtain feed, and the effect of housing systems on the associated behaviours, can help to improve pig welfare. The aim of this study was to identify the key elements controlling feeding behaviour of growing pigs in order to provide a theoretical basis for an individual-based simulation of feeding behaviour and associated welfare issues. Based on empirical data and theories in literature, a conceptual framework was constructed to identify the essential elements affecting feeding behaviour. Behavioural patterns around feeding, including conflict behaviours (e.g. approach or avoidance) and time intervals between feeding, are the result of a complex interaction between various mechanisms that may operate at different levels. To understand how the different mechanisms interact, knowledge of ethology, physiology, psychology, and nutrition on animal, group and housing level is integrated in this framework. Theories of motivation are applied to explain behaviour and to understand how internal and external factors can affect behaviour. These theories state that animals have internal states that motivate an animal to perform certain behavioural patterns. The conceptual framework can be divided into two main processes. The first process is about the formation of feeding motivation and other motivational states in the pigs, the second process concerns decision making i.e. how pigs react behaviourally on the feeding motivation. Important factors affecting feeding motivation and behaviour in this framework are pig characteristics (e.g. gender, weight, growth capacity, coping style, social rank), internal drives (e.g. metabolic energy, nutrient balance, stress level, diurnal rhythm, other motivational states) and external incentives (e.g. temperature, presence of feed, availability feeding place, competition). This interdisciplinary and complex framework shows how individual variation among pigs and social influences interact multidimensional and affect feeding behaviour. The framework is further developed and being tested in an agent-based model to examine effects of interaction between pigs, characteristics of the housing, and management by the farmer on feeding behaviour and associated welfare of pigs.

Sequential sampling: a novel method in farm animal welfare assessment

C.A.E. Heath[1], D.C.J. Main[1], S. Mullan[1], M.J. Haskell[2] and W.J. Browne[1]
[1]University of Bristol, School of Veterinary Sciences, Langford House, BS40 5DU, United Kingdom, [2]SRUC, Roslin Institute Building, EH25 9RG, United Kingdom; cheryl.heath@bristol.ac.uk

Farm animal welfare assessment is used to provide an overall picture of the welfare status of the animals on a farm. It is rarely practical or feasible to assess every animal on the farm, therefore some form of sampling scheme is used where a trade-off is made between time taken and accuracy of the results. Here, how representative the sample results are of the farm as a whole is important, with typically larger sample sizes being associated with greater levels of accuracy but also greater costs in terms of time and resources. Current sampling schemes are generally based on a fixed size sample of animals which may or may not be dependent on the size of the farm. In this talk we will describe sequential sampling, a method commonly used in clinical trials, for reducing sample sizes. In sequential sampling, data is collected sequentially and at stages of the sampling the collected data is assessed as to whether we have enough information to make a decision or whether more needs collecting. This potential early stopping can reduce sample sizes and thus speed up the assessment. This method is suitable for welfare indicators that involve comparing on farm prevalence with a fixed threshold. We illustrate sequential sampling by considering a dataset of lameness prevalence in dairy cows on 80 farms. We compare a fixed size sampling scheme using sample sizes based on the Welfare Quality protocol for dairy cows with a two-stage sequential scheme that has a potential stopping point halfway through sampling. Data has been collected on all cows and farm level prevalence ranges from 8 to 73%, with a mean farm level lameness prevalence of 33%. We use simulation methods to perform simulated welfare assessments for each scheme and then make comparisons in terms of accuracy and time taken. Lameness is scored on a 4 point (1-4) scale with 2 and 3 indicating lame and severely lame cows and there is a correlation (0.58) between the proportion of lame cows and the proportion of these cows that are severely lame. We therefore investigate how we can build the number of severely lame cows into the stopping rule to further improve performance. We end by describing when (and when not) in practice sequential sampling methods could be useful.

Housing and management factors associated with indicators of dairy cattle welfare

M. De Vries[1], E.A.M. Bokkers[1], C.G. Van Reenen[1], B. Engel[2], G. Van Schaik[3], T. Dijkstra[3] and I.J.M. De Boer[1]
[1]Wageningen University, Animal Production Systems group, Elst 1, 6708 WD Wageningen, the Netherlands, [2]Wageningen University, Biometris, Droevendaalsesteeg 1, 6708 PB Wageningen, the Netherlands, [3]GD Animal Health Service, Arnsbergstraat 7, 7418 EZ Deventer, the Netherlands; marion.devries@wur.nl

Knowing synergies and trade-offs between factors influencing different aspects of dairy cattle welfare is essential for farmers who aim to improve the level of welfare in their herds. The aim of this research was to identify and compare housing and management factors associated with the prevalence of lameness, lesions or swellings, dirty hindquarters, and the frequency of displacements in dairy herds in free-stall housing. Seven trained observers collected data regarding housing and management characteristics of 179 Dutch dairy herds (herd size: 22 to 211 cows) in free-stall housing during winter. Lame cows, cows with lesions or swellings, and cows with dirty hindquarters were counted and occurrence of displacements was recorded during 120 min of observation. For each of the four welfare indicators, housing and management factors associated with the welfare indicator were selected in a succession of logistic or log linear regression analyses. Prevalence of lameness was associated with surface of the lying area, summer pasturing, herd biosecurity status, and dry cow groups (P<0.05). Prevalence of lesions or swellings was associated with surface of the lying area, summer pasturing, light intensity in the barn, and days in milk when the maximum amount of concentrates was fed (P<0.05). Prevalence of dirty hindquarters was associated with surface of the lying area, proportion of stalls with faecal contamination, head lunge impediments in stalls, and number of roughage types (P<0.05). Average frequency of displacements was associated with the time of introducing heifers in the lactating group, the use of cow brushes, continuous availability of roughage, floor scraping frequency, herd size, and the proportion cows to stalls (P<0.05). Prevalences of lameness and of lesions or swellings were lower in herds with soft mats or mattresses (odd ratio (OR)=0.66 and 0.58, confidence interval (CI)=0.48-0.91 and 0.39-0.85) or deep bedding (OR=0.48 and 0.48, CI=0.32-0.71 and 0.30-0.77) in stalls, compared with concrete, and in herds with summer pasturing (OR=0.68 and 0.41, CI=0.51-0.90 and 0.27-0.61), compared with zero-grazing. Deep bedding in stalls was negatively associated with prevalence of dirty hindquarters (OR=0.54, CI=0.32-0.92), compared with hard mats. It was concluded that some aspects of housing and management are common protective factors for prevalence of lameness, lesions or swellings, and dirty hindquarters, but not for frequency of displacements.

A cross-sectional study on the prevalence and risk factors for limb and foot lesions in piglets

A.J. Quinn[1,2], A.L. Kilbride[1], L.E. Green[1] and L.A. Boyle[2]
[1]University of Warwick, Life Sciences, Coventry, CV4 7AL, United Kingdom, [2]Teagasc, Pig Development Department, Animal and Grassland Research and Innovation Centre, Moorepark, Fermoy, Co. Cork, N/A, Ireland; laura.boyle@teagasc.ie

A cross sectional survey of 68 integrated Irish pig farms (mean sow herd size: 645) was carried out to determine the prevalence and risk factors for foot and limb lesions in 2948 piglets from 272 litters. One litter was selected per age category 3-7 days (d), 8-14 d, 15-21 d and 22-28 d on each farm and all piglets in each litter were examined for limb abrasions, joint swellings, sole bruising, sole erosion, claw swellings and coronary band injuries which were scored from 0-3 based on size. 58 environmental parameters (crate/pen/flooring dimensions and characteristics [material, slopes, steps, and slat features, etc.]) were recorded for each litter examined. A questionnaire was completed on management, health and performance factors for each farm. The prevalence of each lesion was calculated and multilevel mixed effect logistic regression for 145 factors was carried out using MlwiN 2.27. The prevalence of lesions was: sole bruising=61.5%, sole erosion=34.1%, coronary band injuries=11.3%, limb abrasions=55.7%, swollen joints=2.4% and swollen claws=4.4%. The only risk factors identified as having a significant impact on these lesions were age and the nature of the openings in the floors in the piglet area of the farrowing pen. Age was a risk factor for sole bruising (OR 0.043; CI 0.41, 0.48) and coronary band damage (OR 0.66; CI 0.59, 0.74) (both decreased with age), and skin abrasions (OR 1.16; CI 1.08, 1.24) which increased with age. There was a reduced risk of sole bruising (OR 0.24; CI 0.18, 0.32) and limb abrasions (OR 0.48; CI 0.35, 0.65) in floors with oval shaped openings compared to floors with rectangular shaped openings. Oval shaped openings were predominately associated with plastic coated woven wire floors while rectangular openings were typical of slatted steel floors which are highly abrasive. The low number of risk factors identified was likely related to the uniformity of the farrowing systems used on the farms. Nevertheless, these results lead to the conclusion that steel floors should be avoided for use in farrowing pens to improve piglet limb/foot health.

Pain management in piglets during management procedures: a systematic review and GRADE process

R. Anthony[1], L. Bergamasco[2], J. Coetzee[3], R. Dzikamunhenga[3], S. Gould[3], A. Johnson[3], L. Karriker[3], J. Marchant-Forde[4], G. Martineau[5], J. Mckean[3], S. Millman[3], S. Niekamp[6], A. O'Connor[3], E. Pajor[7], K. Rutherford[8], M. Sprague[3], M. Sutherland[9] and E. Von Borell[10]
[1]University of Alaska Anchorage, Anchorage, AK, USA, [2]Virginia Tech University, Blacksburg, VA 24061, USA, [3]Iowa State University, 1600 S 16th St., Ames, IA 50011, USA, [4]USDA ARS LBRU, West Lafayette, IN 47907, USA, [5]Ecole Nationale Veterinaire, 31076 Toulouse Cedex 3, France, [6]National Pork Board, Des Moines, IA 50325, USA, [7]University of Calgary, Calgary, AB, T2N 1N4, Canada, [8]SRUC, Roslin Institute Building, Easter Bush, Midlothian, EH25 9RG, United Kingdom, [9]AgResearch Limited, Ruakura Research Centre, Hamilton, 3240, New Zealand, [10]Martin Luther University Halle Wittenberg, 06108 Halle, Germany; aoconnor@iastate.edu

Objectives were to review the literature about efficacy of pain mitigation in 1- to 28-day-old piglets undergoing castration, tail docking, teeth clipping, and ear notching or tagging and to develop recommendations for US swine producers for pain mitigation interventions. The project consisted of two aspects: (1) a systematic review of the literature about interventions to mitigate pain in piglets; and (2) a recommendation development using The Grading of Recommendations Assessment, Development and Evaluation (GRADE) process. A review protocol was designed a priori. Data sources were Agricola, CAB Abstracts, PubMed, Web of Science, BIOSIS Previews, and ProQuest Dissertations & Theses Full Text. No restrictions on year of publication or language were placed on the search. Relevant non-English studies were translated. Two reviewers screened citations for relevance and extracted data from relevant articles. From 2,203 retrieved citations, 40 publications containing 52 studies met the eligibility criteria. In 40 studies, piglets underwent castration only and in 7 studies tail docking only. General anesthesia protocols were assessed in 32 trial arms, local anesthetic protocols in 30 trial arms, NSAID protocols in 28 trial arms and 41 trial arms were controls where piglets received placebo or no treatment. There were 45 outcomes extracted from the studies, however only results from studies that assessed cortisol (6 studies), beta-endorphins (1 study), vocalizations (9 studies), and pain-related behaviours (9 studies) were used to make recommendations since other outcomes were reported rarely. Recommendations were developed for 3 interventions (CO_2/O_2 general anesthesia, NSAIDs protocols, and lidocaine). The ability to make strong recommendations was limited by lack of comprehensive reporting of study results and strong certainty about variation in stakeholder values and preferences.

Facilitating technical communication during implementation of the Broiler Directive (2007/43/EC)

A. Butterworth[1], I.C. De Jong[2], C. Keppler[3], L. Stadig[4] and S. Lambton[1]
[1]University of Bristol, Clinical Veterinary Science, Langford, N Somerset, BS40 5DU, United Kingdom, [2]Wageningen UR Livestock Research, P.O. Box 65, 8200 AB Lelystad, the Netherlands, [3]Universität Kassel, Fachgebiet Nutztierethologie und Tierhaltung, Nordbahnhofstrasse 1a, Kassel, Germany, [4]Institute for Agricultural and Fisheries Research (ILVO), Burgemeester Van Gansberghelaan 96, Box 1 9820 Merelbeke, Belgium; andy.butterworth@bris.ac.uk

The Broiler Directive (2007/43/EC) is unique amongst current EU Directives which address Animal Welfare, in that it uses outcome data collected at abattoirs and on farm to monitor on farm broiler welfare and vary the maximum permitted stocking density on farm. In this study we describe how, by bringing together personnel from 22 competent authorities who have responsibility for implementing the directive, and engaging in exchange of information and technical methods regarding the broiler directive, it has been able to identify and quantify differences in approach across EU Member States, and to create a web based technical information sharing portal. The study identified the key components of the measures being assessed by the different member states, and this information is summarised in the paper. Online questionnaires and workshop exercises enabled us to identify priority areas for knowledge transfer and training. For example, food pad dermatitis, hock burn, deaths on arrival and total rejections were identified as measures of medium to low priority in terms of developing knowledge transfer because methods that are already well accepted by competent authorities are available. On the other hand, breast lesions, cellulitis, emaciation, joint lesions, respiratory problems, scratches, wing fractures and a number of environmental measures were identified as having high priority, in terms of developing knowledge transfer. Additionally, workshops resulted in creation of a technical 'sharing' website which provides summary information on the spread of approaches to implementing the Directive, as well as examples of the measures used, how they are assessed, what 'trigger levels' are set, and how judgements are framed and made to enforce to the requirements of the Directive. The study identified that there is significant variability in the stage of implementation between MS, and responses from the participating member states indicated that shared guidance and technical information provided by a web tool may be of value in the set up process for those MS engaged in implementation of the Directive. Summaries of the output from the web tool are presented as examples of the potential for exchange of knowledge and technical material to promote moves toward harmonised application of the Directive across the EU.

Benchmarking stockpeople for welfare assessment

G.J. Coleman, L. Roberts and V. Rohlf
University of Melbourne, Animal Welfare Science Centre, Parkville, 3010, Australia; grahame. coleman@unimelb.edu.au

There are increasing international and local pressures for farm animal welfare monitoring schemes. While animal-based welfare monitoring schemes are likely to improve animal welfare, the impact of such schemes will only be realised by recognising the limitations of stockpeople. According to English *et al.*, the most valuable economic resource of society is its people and stockpeople are the most vital component of animal production systems. Domesticated animals are totally dependent on those who care for them to provide shelter, nutrition and medical treatment if required. Stockpeople are responsible for the welfare of the animals, the quantity as well as quality of the system's output and how efficient and competitive the current system is. The aim of this study was to develop a tool to assess on-farm stockperson performance. Stockperson self-report questionnaires as well as supervisor questionnaires were developed and a protocol to assess stockperson-animal interactions was developed to validate the stockperson questionnaires. A total of 15 farms from four large commercial piggeries comprising a total of 158 stockpersons provided written consent to be involved in the project and 117 stockperson questionnaires, 138 supervisor surveys and 132 behavioural observations were completed. This resulted in 79 complete datasets with matching supervisor reports, stockperson questionnaires and behavioural observations. Twelve stockperson subscales and four supervisor subscales were identified. A number of the stockperson subscales correlated significantly with supervisor assessments or with stockperson behaviour. This provided evidence of validity of the stockperson questionnaire. Test-retest correlations were all significant and indicated acceptable levels of reliability. Following the refinement of the questionnaires, a program, called Prohand Benchmark, was developed in JavaTM as were user manuals for administrators and supervisors. Prohand Benchmark is designed to provide piggery managers/owners with information on how stockpeople perform in comparison with other stockpeople working within the Australian pig industry. In addition to being a benchmarking tool, the program can be used for the purpose of identifying areas for individual stockperson training.

The animal welfare status of EU-imported versus locally produced broiler meat

F. Tuyttens[1], R. Vanderhasselt[1], K. Goethals[2], L. Duchateau[2], J. Federici[3] and C. Molento[3]
[1]Institute for Agricultural and Fisheries Research (ILVO), Animal Sciences Unit, Scheldeweg 68, 9090 Melle, Belgium, [2]Ghent University, Faculty of Veterinary Medicine, Salisburylaan 133, 9820 Merelbeke, Belgium, [3]Federal University of Paraná, Department of Animal Science, R. dos Funcionarios 1540, 80035-050 Curitiba, Brazil; frank.tuyttens@ilvo.vlaanderen.be

Legislative welfare standards for broiler production are stricter in the EU than in many other parts of the world. Although these standards may increase production costs they're not incorporated in World Trade Organisation regulations. The EU broiler industry may therefore be at a competitive disadvantage unless other attributes that differentiate locally produced meat from imported meat influence food purchasing decisions. Animal welfare may be such an attribute as consumers often associate local production with high welfare standards. The aim of this study was to compare broiler welfare on farms in Belgium versus Brazil (the major broiler meat exporter to the EU). Two trained observers, simultaneously and independently, carried out broiler welfare assessments on 11 farms in North Belgium and 11 farms in South Brazil that produced for the EU-market, using the Welfare Quality® protocol. Although the overall welfare category was the same for all farms ('acceptable'), t-tests revealed many country differences at the level of the welfare principles, criteria and measures. Brazilian farms obtained higher scores for three of the four principles: 'Good Feeding' (P=0.007), 'Good Housing' (P<0.001), 'Good Health' (P=0.005). The mean score for the principle 'Appropriate Behaviour' did not differ significantly (P=0.741). Four of the 10 welfare criteria scores were, or tended to be, higher on Brazilian than Belgian farms: 'Absence of Prolonged Thirst' (P<0.001), 'Ease of Movement' (P<0.001), 'Absence of Injuries' (P=0.002), 'Positive Emotional State' (P=0.055). 'Absence of Prolonged Hunger' (P=0.048) and 'Good Human-Animal Relationship' (P=0.002) were the only two criteria with a higher score for the Belgian than Brazilian farms. Although overall broiler welfare was within the 'acceptable' category on all farms, Brazilian flocks obtained more often better scores for various welfare aspects than Belgian flocks did. Assuming that the Welfare Quality® protocol allows a valid quantification of animal welfare, these results suggest that the welfare of broilers produced in Brazil for import to the EU-market is not inferior to that of locally produced broilers. More welfare assessments using animal-based indicators on a larger sample of farms are needed to check whether these findings can be generalised. These results also call for more research to investigate the effect of animal welfare legislation on the actual welfare status on commercial farms.

Assessing laying hen welfare outcomes in the context of the 'good life' resource tiers framework

J. Stokes[1], K. Still[1], J. Edgar[2], M. Cooper[3], J. Wrathall[3] and D. Main[2]
[1]Soil Association, Innovation, South Plaza, Marlborough Street, Bristol, BS1 3NX, United Kingdom, [2]University of Bristol Veterinary School, Animal Welfare and Behaviour Group, Dolberry Building, Langford, Somerset, BS40 5DU, United Kingdom, [3]RSPCA, Farm Animals Department, Wilberforce Way, Southwater, West Sussex, RH13 9RS, United Kingdom; jstokes@soilassociation.org

On-farm assessment of animal welfare requires a combination of animal-based outcomes to monitor the actual state of welfare and resource-based inputs to assess the provision of resources to satisfy welfare. AssureWel, led by University of Bristol, RSPCA and Soil Association (SA), has developed a welfare outcome assessment for laying hens which has become an integrated part of Soil Association Certification (SACL) and RSPCA Freedom Food (FF) farm assurance scheme audits since September 2011. Whilst this assessment predominantly focuses on the presence of negative welfare states such as feather loss and dirtiness, the Farm Animal Welfare Council proposed that, in order to demonstrate positive welfare, the provision of resources 'that an animal does not need for biological fitness but is valued' should be considered. In response a resource tiers framework has been developed which identifies opportunities for a 'good life' that lead to positive welfare outcomes and so are compatible with higher welfare farm assurance schemes. Published evidence and expert opinion were used to define three tiers of resource provision (Welfare +, Welfare ++ and Welfare +++) above those stipulated in UK legislation and codes of practice. The final tiers framework is categorised according to the FAWC's opportunities encompassing mental aspects of welfare; 'Comfort', 'Pleasure', 'Confidence' and 'Interest', and also the additional opportunity of a 'Healthy Life' in order to incorporate an achievable balance between animals being healthy and having 'what they want'. Data on laying hen resource tiers and welfare outcome is being collected across a representative proportion of SA and FF farms and the relationship between resource tier and welfare outcome scores explored. To date, data from 21 farms has been collected, with the plan to gain data from a sample of 60 farms in total. Results of the analysis conducted will be presented, along with an evaluation of the practicality and value of using a resource tiers framework to assess laying hen welfare on farm.

Animal Welfare Quality®: interaction between science and business management

V. Pompe and H. Hopster

Van Hall Larenstein University of Applied Sciences, Animal Welfare Group, P.O. Box 1528, 8901 BV, Leeuwarden, the Netherlands; vincent.pompe@wur.nl

In the Dutch project 'Welfare Monitoring Calves' the interaction between science and business management is one of the key aspects. The project's basic assumption is that the effectiveness of a scientific development depends on the implementation in practice. This indicates that welfare quality system is a co-production of livestock science and veal farmers / veal calf integrators. In the project's Sigma-DMAIC approach of Define, Measure, Analyse, Improve, Control, the co-production between the main developer (science) and the main user (farmer) becomes clear. Both sides produced the definition of the (12) welfare criteria, but science is more responsible for the M+A and farm management more for I+C. From a pragmatic philosophical perspective a theoretical framework will be presented for enhancing the interaction between science and business management by focusing on 'room for manoeuvre' of both sides. The framework comprises the integration of natural science into action research and the conditions regarding the input-throughput-output process. Some outcomes of the workshops with Dutch veal farmers, in which the economic and managerial aspects of welfare improvement were discussed, and the dynamics of setting target values from the welfare criteria will illustrate the framework. In developing a welfare quality system one must understand that science can tell us what worked but not what will work, while management can tell us what will work but not that it worked. It all comes down to: Effect (improved welfare) = Quality (science) × Acceptance (farmers).

Characterization of dairy cattle farms according to welfare level in the Plains of Bogota, Colombia

C. Medrano-Galarza, A. Zuñiga-López and F.E. García-Castro
*Corpoica, Animal Health and Welfare, Km 14 Via Bogota, Mosquera, Colombia;
fgarciac@corpoica.org.co*

Assessing welfare at the farm level can be used as a tool for decision-making and for implementing improvement plans by producers. Assessments can also be used as a source of information for establishing food quality insurance schemes and legislation. Integral welfare assessment protocols should involve aspects related to animal-based, management-based and resource based indicators. In Colombia, dairy farms are mainly pasture-based. However, to have cows on pasture does not necessarily mean good welfare; there are many factors that can affect the quality of life of dairy cows in extensive systems. The objective of this study was to evaluate the level of welfare on dairy cattle farms in the Plains of Bogota, Colombia. An observational study was done on 25 farms (two visits per farm; summer and winter 2013). We evaluated 27 variables based on the 'Five Freedoms'; 14 animal-based and 13 resource-based. Overall, 48% of the farms received a 'Good' welfare classification and 52% were classified as 'Fair'. Twelve per cent of the farms showed more than 15% of the animals were assessed to be outside acceptable ranks for body condition. Severe injuries on the underbelly, due to flies, where found on 4% of the farms. At 28% and 20% of the farms more than 15% of the animals presented fly counts over 150 (*Haematobia irritans*) and 25 (*Stomoxys calcitrans*), respectively. The mean prevalence of clinical and subclinical mastitis was 1.3% and 31.2%, respectively. However, 12% and 84% of the farms were over the accepted prevalence limits for clinical (<3%) and subclinical mastitis (<15%). Sixteen per cent of the farms had a mean prevalence of lameness over 10%. The mean flight distance was 2.6±1.2 meters (min=0; max=9.2). Only 8% of the farms had more than 15% of the animals with flight distance above 5 meters. The behaviour of lactating cows during milking was mainly calm (90.9% of all evaluated animals; 6.6%: agitated; 2%: uneasy; and 0.6%: aggressive). No farm had more than 15% of their cows categorized as uneasy or aggressive. The cleanliness score of the water trough most frequently found was partially dirty (44% of the farms), 24% scored dirty. Determining the level of welfare at the farm level encourages producers to find ways to make improvements, turning weaknesses into strengths in striving for excellence. This study is the first of this kind in Colombia, and could be used as a guide to promote discussions about the importance of welfare assessment schemes on farms.

The development of practical measures to benchmark pig welfare in the Australian pork industry

L. Hemsworth and J. Skuse
Animal Welfare Science Centre, The University of Melbourne, Parkville, 3010, Australia;
lauren.hemsworth@unimelb.edu.au

The Australian Pork Industry Quality Assurance Program requires producers to be annually audited against a set of standards and performance indicators, but does not provide the opportunity to monitor the welfare status of pigs over time and/or benchmark the welfare status of animals between farms. This project aimed to identify and examine the suitability of a range of valid, reliable, and feasible animal-based welfare indices for inclusion in a practical on-farm pig welfare benchmarking tool to be applied by Australian producers for self-auditing purposes. Furthermore, this protocol will be applicable to sows and growing/finishing pigs, across all stages of production and within all forms of Australian production system. An extensive review of the relevant literature identified key animal-based welfare indices, and the most relevant measures of pig welfare were selected by a focus group comprising pertinent research and industry personnel. The practicality of these validated indices for on-farm assessment was examined at a range of production systems representative of the Australian Pork Industry. The validated animal-based welfare indices recommended for inclusion in an on-farm pig welfare benchmarking tool are body condition score, body lesion score, vulva lesion score (sows only), tail lesion score (growing/finishing pigs only), bursitis, lameness score, coughing, sneezing, stereotypic behaviour (sows only), and morbidity. Methodology modifications necessary for on-farm assessment in Australian conditions are recommended for a more simplified and practical scoring of body condition, body lesions, and lameness. The validity and reliability of the recommended welfare indices and on-farm methodology modifications are currently being evaluated via comparisons with previously validated measures of body condition, body lesions and lameness; in order to fully develop an effective on-farm pig welfare benchmarking tool suitable for the Australian Pork Industry. An on-farm pig welfare benchmarking tool, which provides a simple, validated set of pig welfare indicators, would enable producers and the industry to monitor pig welfare over time, to demonstrate improvement in animal welfare outcomes over time, to identify areas of improvement for pig welfare, and to compare pig welfare across units in multisite enterprises. In addition, the proposed benchmarking tool will also ensure that the Australian Pork Industry continues to remain at the forefront of International developments in animal welfare, maintain a high level of ethical standards and promote sound welfare practices.

Implementation of the Welfare Quality® broiler monitor on Dutch broiler farms

I.C. De Jong, H. Gunnink and V. Hindle
Wageningen UR, Livestock Research, P.O. Box 65, 8200 AB Lelystad, the Netherlands;
ingrid.dejong@wur.nl

A project was started to implement the Welfare Quality[*] broiler monitor on Dutch broiler farms. Aims of the project were (1) to study if the monitor could be simplified resulting in a less time-consuming on-farm assessment, and (2) to study if the monitor could be used as a management tool for farmers and veterinarians. Fifty farms participated in the project, either with standard, fast growing broilers (n=38) or slower growing birds in alternative systems (n=12). Per farm, one house was monitored twice in 2013 and will be monitored twice in 2014. Data of the first two visits were used to calculate Spearman rank correlations between individual measures as the basis for simplification. Three simplification strategies were defined in a previous study and tested in the current study: (1) prediction of gait-scores on-farm from measures of hock-burn (HB) on-farm; (2) prediction of gait score on-farm from HB measures at slaughter; and (3) prediction of clinical scores on-farm from measures of footpad dermatitis (FPD) and HB at slaughter. Analysis of correlations between gait score on-farm and HB on-farm (strategy 1) and at slaughter (strategy 2) showed that significant correlations were absent for flocks with standard broilers and these strategies were not further analysed. A significant high correlation was found between measurements of FPD and HB on farm and at slaughter (fast growing broiler flocks, severe FPD: R_{sp}=0.79, P<0.001; severe HB: $R_{s}p$=0.48, P<0.001; flocks with slower growing strains, severe FPD: R_{sp}=0.64, P<0.001, severe HB: R_{sp}=0.35, P<0.10). Although correlations between FPD and HB at slaughter and cleanliness measured on-farm were lower (fast growing strains: R_{sp}=0.40, P<0.01 (severe FPD – cleanliness score ≥2) and R_{sp}=0.32, P<0.01 (HB – cleanliness ≥2); slower growing strains: R_{sp}=0.12 and R_{sp}=0.21, (ns) for cleanliness ≥2 – severe FPD and HB respectively), it was decided to further analyse this simplification strategy. Predicting clinical scores on-farm from measures of HB and FPD at slaughter affects scores for principle 2 (good housing), criterion 3 (comfort around resting), principle 3 (good health) and criterion 6 (absence of injuries). Scores for these principles and criteria were calculated using the full model and the simplified model. Sensitivity and specificity of the simplified model was determined and Spearman rank correlations were determined between the full and simplified model for each criterion and principle affected by the simplification. Agreement between the full and simplified model was high (R_{sp}=0.92 at least). It was decided to use the simplified model for the farm visits in 2014. In 2014 the project will focus on the monitor as a tool for the farmer to improve the management with respect to broiler welfare.

Economic analysis of a management tool for the improvement of laying hen health on organic farms

C. Keppler[1], D. Moeller[2] and U. Knierim[1]
[1]University of Kassel, Farm Animal Behaviour and Husbandry Section, Nordbahnhofstr. 1a, 37213 Witzenhausen, Germany, [2]University of Kassel, Farm Management Section, Steinstrasse 19, 37213 Witzenhausen, Germany; ckeppler@wiz.uni-kassel.de

A management tool was developed and optimized comprising a recording and assessment protocol regarding animal health and laying performance as well as structured forms for feedback, including benchmarking, and decision support for advisory purposes. It was applied and tested on 11 laying hen and 6 rearing farms over two rearing and laying periods by trained advisors. Costs of application of the management tool and implementation of advisory recommendations were analysed by comparing actual and simulated cost-performance calculations. On the laying hen farms predominantly breast bone deviations (10 farms), incomplete feathering and sometimes lesions due to feather pecking and cannibalism (4 farms), and low body weight or flock uniformity (11 farms) were found. Risk factors were mainly identified in the areas of process management, care during change from rear to laying and feeding management. Consequently, improved process management, increased control frequencies, provision of ascending aids early in the laying period, feed analyses and feeding advice were among the recommended measures. Actual costs per egg (without marketing) on the project farms were between 15.30 and 24.23 cent, with a substantial cost degression with increasing farm size. The management tool application would cost on average 1.35 cent/egg (2.81-0.59 cent/egg on farms with about 1,300 or 8,000 to 9,000 hens, respectively). In case of animal health problems, the simulated improvement measures would cause additional costs of 0.00 to 3.50 cent/egg when taking into account potential improvements in laying performance. The break-even point on average would be at 20.45 cent/egg (26.7-18.5 cent/egg). Farmers and advisors evaluated the management tool mostly to be practicable and helpful. Through its broad approach and scientific foundation the tool provides a novel way to transfer current knowledge regarding health management into practice. It appears that an integration into a quality assurance system is possible. However, additional costs may arise which should be compensated by higher egg prices, especially for small farms.

Effective communication of on-farm welfare assessment results on non-caged laying hen farms

S. Butcher[1], M. Fernyhough[1], E. Heathcote[1], D.C.J. Main[2], K. Still[3] and J. Stokes[3]
[1]RSPCA, Wilberforce Way, Southwater, Horsham, West Sussex, RH139RS, United Kingdom, [2]University of Bristol Veterinary School, Langford House, Langford, Somerset, BS405DU, United Kingdom, [3]Soil Association, South Plaza, Marlborough Street, Bristol, BS13NX, United Kingdom; sophie.butcher@rspca.org.uk

AssureWel, a project led by University of Bristol, RSPCA and Soil Association, has developed protocols to assess farm animal welfare in farm assurance schemes. Implementation of the laying hen protocol within the RSPCA Freedom Food and Soil Association schemes (Sept 2011-Sept 2013) has demonstrated significant improvement in some welfare measures. Effective communication to motivate producers to respond positively to their assessment results is considered a key driver behind these improvements. Effective communication of assessment results aims to maximise producer engagement, increase issue awareness and foster behavioural change. A number of tried and tested tools have been developed by AssureWel to maximise these impacts: (1) Benchmarking compares a producer's scores to their peers'. As such, benchmarking demonstrates what is achievable, can provide targets and prompt responsive action; (2) Threshold levels communicate acceptable levels of critical outcome measures and ensure clear communication of expectation; (3) Joint scoring between the assessor and producer promotes dialogue and discussion, which can improve producer knowledge and awareness and facilitate the uptake of self-assessment; (4) Motivational techniques such as open questioning and reflective listening are employed by the assessor to enhance dialogue with producers. Poor assessment results are presented as opportunities for improvement; (5) Post assessment advice and support: provision of published materials, producer-focused websites and dedicated Welfare Advisors support producers in their improvement targets. Using such approaches to communicate welfare assessment results effectively on-farm encourages producers to take positive action post-assessment in the form of husbandry and/or resource change. Analysis of data collected on 459 farms from Nov 2012 – Nov 2013 show a total of 846 changes were made, ranging from 1-5 changes per farm, affecting 7,865,423 birds. Producer changes included improvements to the range (29%), additional enrichment provision (28%), disease management (28%) and maintenance of litter condition (13%). Categories are ranked by their relationship with risk reduction of welfare issues. As reported by Mullan, year-on-year data shows significant improvement in some welfare measures with feather loss reduced by a third. Implementation of factors described above may be one contributing factor ascribed to welfare improvement on laying hen farms.

To inspect, to motivate or to do both: a dilemma for on-farm inspection of animal welfare

I. Anneberg[1], M. Vaarst[1] and P. Sandøe[2]

[1]Aarhus University, Department of Animal Science, Blichers allé 20, 8830 Tjele, Denmark, [2]University of Copenhagen, Department of Large Animal Sciences / Department of Food and Resource Economics, Rolighedsvej 25, 1958 Frederiksberg, Denmark; inger.anneberg@agrsci.dk

In European countries, authorities try to enforce animal welfare legislation through inspections followed up by warnings or penalties in instances where a lack of compliance is found. However, the fairness and efficiency, and ultimately the public acceptance of the system, critically depend on the performance of individual inspectors. On the one hand it may be argued that the controllers should check only whether farmers comply or not with animal welfare regulations. Here, the key-value is the rule of the law, and that all offenders should be treated equally. On the other hand, it may be argued that an important component of inspection is to enter into dialogue with the farmers. This may be based on a more forward-looking view aimed at motivating farmers to look after the welfare of the animals in their care. This paper presents results from an interview-study into how animal welfare inspectors from the Danish authorities view their own role and tasks when carrying out welfare inspections. Such inspections are carried out for regulatory purposes on a number of Danish livestock farms per year. During the unannounced visit, the inspector checks all the provisions of the animal welfare legislation, most of which relate to the physical environment of the animals. 22 unannounced inspections on farms of different size were observed. The observations were followed by qualitative interviews with 12 inspectors (and 12 farmers), chosen from the inspections on selected large pigs- and dairy cattle farms. In the main results, a theme of disagreement presented itself and revealed different attitudes in terms of the possibility of engaging in a dialogue with the farmers. The first theme focused on the preventive aspect. Here it was seen as meaningful for the inspectors work to talk with the farmers and inform about the background for the legislation The second theme had its focus on compliance and on the avoidance of engaging in dialogue with the farmer regarding the reasons for the regulations. Moreover, a theme of agreement showed interpretation of the legislation as unavoidable. Also, a theme of seeing the inspectors as the protector of the individual animal came forward. Besides, inspectors found it necessary that the authorities secure time for them to meet and calibrate their findings – and finally a theme of seeing the EU cross compliance as bringing complications into the communication with the farmers also came forward. The research points towards an important perspective of bringing more attention to the communication during these inspections.

Investigating the humaneness of two novel mechanical killing devices for killing poultry on farm

J.E. Hopkins[1,2], D.E.F. Mckeegan[1], J. Sparrey[3] and V. Sandilands[2]
[1]University of Glasgow, College of Medical, Veterinary & Life Sciences, University of Glasgow, Bearsden Road, G61 1QH, United Kingdom, [2]SRUC, Animal and Veterinary Sciences Group, SRUC, West Mains Road, Edinburgh, EH26 0PH, United Kingdom, [3]Livetec Systems Ltd, Building 52, Wrest Park, Silsoe, Bedford, MK45 4HS, United Kingdom; jessica.hopkins@sruc.ac.uk

Worldwide it can be estimated that a maximum 9.1 billion birds may need to be killed on farm each year. The method in which these birds are killed on farm is crucial to poultry welfare. Two novel mechanical poultry killing devices: Modified Rabbit Zinger (MZIN), a novel mechanical cervical dislocation device (NMCD) and a control (manual cervical dislocation – MCD) were tested on 180 birds across two bird types and ages (broilers/layers × juveniles/adults). Efficacy of devices was determined in two ways: (1) latencies of reflexes post treatment application; and (2) post mortem analysis (device success – desired anatomical effect on bird is observed and resulted in death; and kill success – adequate anatomical damage achieved in order to kill bird). Failed kill attempts were immediately emergency euthanized and reflex/behaviour data was not recorded (17/180 birds). Kill success ($F=19.96_{(2,167)}$, P<0.001) and device success ($F=7.33_{(2,167)}$, P<0.001) were significantly affected by kill treatment through GLMM analysis, with NMCD and MCD achieving 100.0±0.0% kill success rate and the MZIN achieving 71.7±5.9%. Device success rates were NMCD=41.7±6.4%; MZIN=70.0±6.0%; and MCD=26.7±5.8%.Only device success was significantly affected by bird type ($F=9.55_{(2,163)}$, P=0.002), with greater device success in broilers. Mean maximum duration times for key reflexes related to consciousness and brain death were for jaw tone (NMCD=8.8±1.3 s; MZIN=4.8±1.1 s; MCD=6.8±1.3 s), for nictitating membrane (NMCD=1.8±0.7 s; MZIN=4.0±1.2 s; MCD=3.3±1.4 s), for pupillary (NMCD=41.5±2.2 s; MZIN=8.3±1.7 s; MCD=41.0±2.2 s) and rhythmic breathing (NMCD=0.0±0.0 s; MZIN=6.5±1.6 s; MCD=0.0±0.0 s). Jaw tone ($F=4.26_{(2,167)}$, P=0.016), pupillary ($F=9.49_{(2,167)}$, P<0.001) and rhythmic breathing ($F=4.16_{(2,167)}$, P=0.017) were significantly affected by kill treatment. Post mortem analysis showed that NMCD and MCD achieved 100% cervical dislocation for all birds. There was a significant difference in number of carotid arteries severed ($F=4.54_{(2,106)}$, P=0.035) between NMCD and MCD, with 71.7 and 58.3% ≥1 carotid arteries severed respectively. It was concluded that the NMCD was the best mechanical killing device in comparison to MZIN and also performed better than MCD, therefore will be taken forward for on farm user-reliability testing.

Environmental factors associated with success rates of Australian stock herding dogs

E.R. Arnott, J.B. Early, C.M. Wade and P.D. Mcgreevy
The University of Sydney, Faculty of Veterinary Science, The University of Sydney, New South Wales 2006, Australia; elizabeth.arnott@sydney.edu.au

Australia has 91,000 livestock producers, who each employ an average of three to four working dogs. The approach to the production and usage of these dogs has lacked a scientific basis as research in the field has been very limited. An average of 20-25% of working dogs recruited for training in Australia are culled due to a lack of suitability. Identifying factors associated with stock herding dog success and failure will enable producers to adapt their practices to gain maximum financial return from their dogs as well as respond to the growing public awareness of the welfare issues associated with food production. This study aimed to investigate the current management practices associated with stock herding dogs on Australian farms. A parallel goal was to determine whether these practices and the characteristics of the dog handlers were associated with success rates (the percentage of dogs acquired for work that ultimately become successful working dogs). Data on a total of 4,027 dogs were acquired through The Farm Dog Survey which gathered information from 812 herding dog owners around Australia. Regression analysis was used to identify significant associations ($P<0.05$) between the respondents' reported success rate and 33 variables that related to handler and dog characteristics and husbandry and training techniques. The following factors were found to be significantly associated with success rate: dog breed ($P=0.027$), method of housing dogs ($P<0.001$), participation in dog trials ($P=0.034$), the insurance status of the dogs ($P=0.04$), the age ($P=0.002$) and training level ($P=0.03$) of the dog at acquisition, the use of electric collars ($P=0.001$), use of positive reinforcement ($P=0.01$), exercise frequency ($P=0.003$), the handler's view of their dogs ($P=0.006$), handler conscientiousness personality score ($P=0.007$) and the expense handler's estimate they would undertake to treat their best dog for an illness to return it to work ($P<0.001$). A number of management practices and human traits were associated with canine outcomes. The significance of husbandry and training techniques suggests the importance of addressing working dog welfare standards. Several other significant variables infer a need to foster the canine-human bond to optimise success. These findings demand recognition of the role the dog handler has in influencing results and provide a guide for areas of further investigation for optimising care and management of Australian stock herding dogs. The insights also have potential relevance to laboratory dogs and other working dog sectors.

Assessing the effects of a disease control programme on dairy goat welfare

K. Muri[1], N. Leine[2] and P.S. Valle[3]
[1]Norwegian University of Life Sciences, Faculty of Veterinary Medicine and Biosciences, Dept. of Production Animal Clinical Sciences, P.O. Box 8146 Dep, 0033 Oslo, Norway, [2]Vennisvegen 950, 2975 Vang i Valdres, Norway, [3]Kontali Analyse AS, Industriveien 18, 6517 Kristiansund, Norway; karianne.muri@nmbu.no

The Norwegian dairy goat industry has largely succeeded in controlling caprine arthritis encephalitis (CAE), caseous lymphadenitis (CLA) and paratuberculosis through a voluntary disease eradication programme called Healthier Goats (HG). The aim of this study was to apply an on-farm welfare assessment protocol to assess the effects of HG on goat welfare. The protocol incorporated both resource- and animal-based welfare indicators, including a preliminary version of qualitative behavioural assessments with five prefixed terms. Twenty goats in each herd were randomly selected for observations of human-animal interactions and physical health. The latter included registering abnormalities of eyes, nostrils, ears, skin, lymph nodes, joints, udder, claws and body condition score. Thirty dairy goat farms were visited, of which 15 had completed disease eradication and 15 had not yet started. Three trained observers assessed the welfare on ten farms each. For individual-level data, robust clustered logistic regression analyses with farm as cluster variable were conducted to assess the association with disease eradication. Wilcoxon rank-sum tests were used for comparisons of herd-level data between the two groups. The qualitative behavioural terms were analysed individually. Goats with swollen joints (indicative of CAE) and enlarged lymph nodes (indicative of CLA) were registered on 53% and 93% of the non-HG farms, respectively, but none of the HG farms. The only other health variables with significantly lower levels in HG herds were skin lesions (P=0.008) and damaged ears due to torn out ear tags (P=0.000). Goats on HG farms showed less fear of unknown humans (P=0.013), and the qualitative behavioural assessments indicated that these herds were calmer than non-HG herds. Significantly more space and lower gas concentrations reflected the upgrading of buildings usually done on HG farms. In conclusion, HG has resulted in some welfare improvements beyond the elimination of infectious diseases. The protocol was considered a useful tool, but larger sample sizes would increase the reliability of prevalence estimates for less common conditions, and increase the power to detect significant differences between groups. Despite the obvious link between disease and suffering, this aspect is rarely taken into account in the evaluation of disease control programmes. We therefore propose that welfare assessment protocols should be applied to evaluate the merits of disease control programmes in terms of animal welfare.

Stress physiology and welfare of fish in aquaculture

E. Sandblom[1], A. Gräns[1], B. Algers[2], C. Berg[2], T. Lundh[2], M. Axelsson[1], L. Niklasson[1], B. Djordjevic[3], H. Sundh[1], H. Seth[1], K. Sundell[1], J.A. Lines[4] and A. Kiessling[2]

[1]University of Gothenburg, Box 463, 40530 Göteborg, Sweden, [2]Swedish University of Agricultural Sciences, Box 234, 53223 Skara, Sweden, [3]Norwegian University of Life Sciences, Box 5003, 1432 Ås, Norway, [4]Silsoe Livestock Systems, Wrest Park, MK45 4HR Bedford, United Kingdom; albin.grans@bioenv.gu.se

Aquaculture is the fastest growing food producing sector. Yet, many challenges and questions remain on how the welfare of fish can be defined, measured and ensured. While the industry requests ethically acceptable methods for handling and stunning fish prior to slaughter, little is known about how fish perceive and respond physiologically to such interferences. CO_2 is currently used to stun fish in aquaculture, but lately this method has been questioned and electric exposure has been proposed as a more ethical alternative. In the lab, we quantified physiological stress responses to these methods in cannulated Arctic char by measuring blood pressure, heart rate and ventilation whilst also collecting blood samples for cortisol analysis. The lab-based studies were complemented with an on-site study where behaviours and blood samples for cortisol analysis were obtained from Arctic char prior to slaughter in a facility running both stunning methods in parallel. The lab results confirmed that CO_2 exposure results in profound behavioural and physiological stress responses, and neither temperature reduction nor additional O_2 eases this situation. Electric exposure for 30 s resulted in a marked blood pressure increase, followed by ventilatory arrest that eventually killed the fish from cardiac ischemia. With a shorter 5 s electric exposure the ventilatory arrest was reversible, but signs of systemic stress responses, including hypertension and increased plasma cortisol levels were evident. In the on-site study, CO_2 exposure triggered aversive struggling and escape responses for 5-10 min before the fish was immobilized, while fish exposed to an electric current were instantly immobilized. On average, it took 5 min for the fish to recover from an electrical stunning, whereas fish stunned with CO_2 failed to recover. Electrically stunned fish had more than twice as high levels of plasma cortisol compared to fish stunned with CO_2. This result is surprising considering that the behavioural reactions were much more pronounced following CO_2 exposure. It is suggested that the relationship and timing of ventilatory failure and loss of consciousness following electrical stunning needs further study to ensure that welfare is not compromised. Attempts are now being made to assess stress physiology and welfare in fish using biotelemetry techniques allowing us to monitor focal animals throughout different production phases in aquaculture settings.

Enrichment and socialisation of the Göttingen Minipig

H. Lorentsen
Ellegaard Göttingen Minipigs A/S, Soroe Landevej 302, 4261 Dalmose, Denmark; hl@minipigs.dk

Enrichment and socialization of the Göttingen Minipig The Göttingen Minipigs are bred in Denmark and used in the field of animal experimentation. Minipigs are curious, social and highly intelligent animals. They benefit from different kinds of stimulation and an environment that respond to their needs. Contact with conspecifics as well as staff is crucial for their wellbeing. Likewise they should be offered different kinds of materials to ensure their urge to root and chew. Even with small changes in the environment it is possible to improve the welfare of the minipigs and reduce stress. Reduction of the level of stress will make them easier to handle and increase the validity of study results. Knowledge of the normal behaviour and nature of the minipigs is important – but what is normal behaviour of a minipig? What is going on in the mind of a minipig? This presentation will include lots of photos and videoclips of minipigs in action. It is shown how minipigs behave in different situations. The aim is to show the audience that minipigs are active animals. With proper stimulation and training they are likely to interact with staff in a positive way to ensure quality in the handling, dosing and sampling procedures.

Welfare issues during traditional slaughter of goats in Pretoria, South Africa

D.N. Qekwana[1], C.M.E. Mccrindle[1], J. Oguttu[2] and B. Cenci-Goga[3]
[1] *Faculty of Veterinary Science, Section Veterinary Public Health, Department of Paraclinical Science, University of Pretoria, Private Bag X04, 0110 Onderstepoort, South Africa,* [2]*Department of Agriculture and Animal Health, College of Agriculture and Environmental Science, University of South Africa, Private Bag X11, 1710 Florida, South Africa,* [3]*Università degli Studi di Perugia, Dipartimento di Medicina Veterinaria, Via San Costanzo, 06121 Perugia, Italy; nenene.qekwana@gmail.com*

Traditional slaughter of goats in South Africa is performed to venerate the ancestors, address personal problems for celebration of marriage and birth as well as during funerals. The practice is permitted by meat safety legislation and the Constitution, provided the meat is not sold but used for personal consumption during the ceremony. The objective of the study was to assess animal welfare issues associated traditional slaughter of goats among communities living in and around Pretoria. Participatory research methods including structured interviews were utilised. Respondents (n=100) who had been involved in traditional slaughter were interviewed. In addition, homes of four of the respondents interviewed were visited to observe the slaughter process. Goats were transported after purchase by vehicle (45.81%) on foot (29.79%), or (24.4%) by trailer, bus, truck or taxi. The distance travelled (67.61%) was usually <10 km, and in all cases <50 km. The most common method of restraining goats during transport (56.82%) and slaughter(66.29%) was by tying all four legs a together. Other methods of restraint included a rope tied around the neck of the goat(18.18%) or placing it in a sack (1.14%). At the place of slaughter, 29.21% are held in a kraal (pen) before slaughter, while 3.37% are slaughtered immediately on arrival and 26.76% slaughtered within an hour. In total, 97.34% of the goats were slaughtered with 24 hours. No stunning was used and the throats were slit while the goats were restrained. It was concluded that there is a need to research practical ways to address animal welfare issues during traditional slaughter taking into consideration people's cultural norms and religious beliefs.

Poster 1

Animal Pain: Perception, evaluation and medical care by veterinary clinicians in the region of Dakar

K.J.F. Adje[1], P.S. Kone[2] and S.N. Bakou[3]
[1]Institut Pasteur de Côte d'Ivoire, Environnement et santé, 08 BP 1883 Abidjan 08, 225, Cote d Ivoire, [2]Ecole Inter-Etats des sciences et de Médecine Vétérinaire de Dakar, Sénégal, Environnemnt et santé, BP 5077, Dakar 221, Senegal, [3]Ecole Inter-Etats des sciences et de Médecine Vétérinaire de Dakar, Sénégal, Santé et Production animale, BP 5077, Dakar 221, Senegal; akjf17@yahoo.fr

In recent years, support for animal pain is a growing place in the Veterinary Clinic. The aim of this work was to evaluate the perception and the medical care of animal pain by clinical veterinarians in the region of Dakar (Senegal). Our area of work was the Dakar region. Our work is a cross-sectional descriptive study. The ground survey was held from March 1 to April 31, 2012. The questionnaire is done through direct interview. To facilitate the collaboration of the interviewed a letter of intent has been sent to them prior. Finally, the opinion of respondents for each statement has been determined using the Likert scale. The collected data was made using Epidata© 3.1. Excel software was used for the descriptive analysis. In total, total 20 clinicians have been surveyed. The majority of respondents (17/20) intervened in all stockfarming. The study population was dominated by men (15/20). Most of the respondents (19/20) has claimed to have heard once talk about animal welfare. Among the five (05) principles of animal welfare only two (02) are known to the clinician. All have said that animals are able to feel pain. Twelve clinicians associated behavioural and injury criteria to recognize pain in animals. More than half of clinicians (11/20) indicated that they use in first intention to ketamine for anaesthesia. After heavy surgery, 50% of clinicians said implement analgesia during the first 24 hours following the intervention. Costs related to the treatment of animal pain can be evaluated according to the respondents, 31 EUR. Relating to animal welfare knowledge levels are low. Implement analgesia after surgery is encouraging.

Poster 2

Evaluation of education of veterinarians in animal welfare issues

S.M. Huertas[1], D. César[2], J. Piaggio[1] and A. Gil[1]
[1]Universidad de la Republica, Facultad de Veterinaria, Lasplaces 1550, 11600, Uruguay, [2]Instituto Plan Agropecuario, Br. Artigas 3802, 11100, Uruguay; stellamaris32@hotmail.com

Training and education of veterinarians in animal welfare issues is the key to improve the health and welfare of food production animals, as well as to contribute increasing safety and quality of food products. A consultation process to veterinarians about training and knowledge received on animal welfare issues has been carried out in Uruguay during March-April 2012. A short questionnaire was sent by email to 250 veterinarians listed on the database of the Continuing Education Program of the Veterinary School of the University of the Republic. The questionnaire asked about the number of animal welfare courses attended, their main field of work, and the importance given to the information provided and the topic addressed. From the 250 veterinarians surveyed, 112 (45%) answered the questionnaire. Of those who answered, 47% reported having attended to three or more animal welfare courses, 18% (20) to two courses and 22% (25) to one. The 82% of respondents work with large animals (beef and dairy cattle) or in the meat processing industry equally. 10% works in areas related to small animals (especially pets) and poultry (production and industry). The 95% of respondents said they felt that the welfare of animals has a very high significance in their daily work with animals. In relation to the applicability of animal welfare concepts in their job, 65% reported 'a lot'. 85% considered that animal welfare should be part of the veterinary curricula and all of them expressed their willingness to attend again to other training courses on the subject if they were offered. These results highlight the importance afforded by Uruguayan veterinarians to improve their knowledge on animal welfare, the need to include specific courses in the veterinary curricula and to work developing continuing education programs in animal welfare issues for veterinarians.

Poster 3

Doing justice to the nature of the animal. Reformational philosophy and the animal welfare debate

C.J. Rademaker[1], G. Glas[1] and H. Jochemsen[2]
[1]VU University, Faculty of Philosophy, De Boelelaan 1105, 1081 HV Amsterdam, the Netherlands;
[2]Wageningen University and Research Centre, Communication, Philosophy and Technology: Centre for Integrative Development, Hollandseweg 1, 6706 KN Wageningen, the Netherlands; corne.rademaker@gmail.com

Any attempt to assess animal welfare needs to face the question what animal welfare is to be understood. Currently, three dimensions are deemed important in animal welfare theory: biological functioning, affective states, and natural living. However, systematic reflection as to how these three dimensions can possibly be combined into an integrated and coherent whole is lacking. In this contribution we therefore investigate whether insights from Reformational philosophy can provide for the desired integration and coherence. We start out with an overview of the different positions that are taken in the animal welfare debate and the problems associated with each of them. Subsequently, we introduce the characteristics of Reformational philosophy and investigate whether Reformational philosophy can integrate the different animal welfare dimensions into an integrated and coherent whole. It turns out that the anti-reductionist nature of Reformational philosophy is highly valuable in the animal welfare debate. By distinguishing between a physical, biotic, and psychic substructure in the animal as a whole, the biological functioning and affective states conceptions of animal welfare can be related to the biotic and psychic substructure, respectively. Because the natural living conception is found to be too static to properly account for the dynamics of reality, this conception of animal welfare is reinterpreted as approaching the animal as a whole, both to its typical, structural and unique, individual side. Moreover, the affective states conception was reformulated as psychic functioning to emphasize that there is always an observable component to the affective state of the animal. Using scientific findings, it is argued that the animal needs an 'appropriate challenge' of environmental stimuli for good biotic and psychic functioning. This also provides a starting-point for animal welfare assessment. In conclusion, Reformational philosophy can provide for a framework in which important intuitions from each of the animal welfare dimensions can be combined into an integrated and coherent whole. For the empirical sciences this means that interdisciplinary research on animal welfare ought to be encouraged.

Poster 4

Farmed, laboratory, companion and wild animals: equal welfare assessment for all?

P. Hawkins
RSPCA, Research Animals Department, Wilberforce Way, RH13 9RS, United Kingdom;
penny.hawkins@rspca.org.uk

The Royal Society for the Prevention of Cruelty to Animals (RSPCA) has a strong interest in welfare assessment, because understanding indicators of animal suffering is an essential component of the Society's work to prevent pain or distress. Work on welfare assessment is undertaken by the RSPCA Science Group, which comprises the Farm, Laboratory, Companion and Wild Animal Departments, and by the Society's Veterinary Department and RSPCA International. The Society is working to review and consolidate its approaches to welfare assessment in all of these areas, according to the premise that the needs of individuals of any given species are essentially the same – regardless of the context in which they are used by, or interact with, humans. However, there are both similarities and differences between the approaches to welfare assessment of the four RSPCA Scientific Departments. For all areas of human-animal interaction, the RSPCA takes the approach of basing welfare assessment primarily on outcomes, in the form of behavioural and physiological indicators of suffering or of wellbeing. Inputs, i.e. what was provided for or done to the animal, are taken into account but are not the primary basis. The need for structured, objective protocols or systems for observing and monitoring animals is also a constant for farmed, laboratory, companion and wild animals. Properly structured, objective systems help to promote consistency between different observers, and are also helpful when training people to observe animals and record what they see. Despite these overarching similarities, the approach to welfare assessment for animals in different situations varies with a number of factors including: (1) the number of animals to be assessed; (2) the resource available to support monitoring programmes; (3) whether suffering is expected; (4) views and beliefs among the (external or internal) target audiences; (5) the legislative framework and guidelines applying to each context of human-animal interaction. The presentation will further explain and discuss these factors, and set out how the RSPCA's Science Group is working to achieve a more cross-disciplinary approach with the ideal of attaining 'equal welfare assessment for all'. Discussion would be welcome as to whether this is a realistic goal – and if not, how any obstacles might be overcome.

Poster 5

Description of some handling practices of beef cattle by farmers related to animal welfare

S.M. Huertas[1], A. Gil[1], D. César[2], J.V. Alsúa[3] and J. Piaggio[1]
[1]Universidad de la Republica, Facultad de Vetrinaria, Lasplaces 1550, 11600, Uruguay, [2]Instituto Plan Agropecuario, Br. Artigas 3802, 11100, Uruguay, [3]Private veterinary practitioner, Bella Union 516, Salto, Uruguay; stellamaris32@hotmail.com

In Uruguay, a South American country with 12 million heads of beef cattle and 3 million people, animal welfare is a national issue. For several years many institutions related to animal production have been working together in order to increase awareness of animal welfare and improve animal husbandry practices. To get information about these practices and to know the reality at farm level at the north of the country, an in-person survey to 300 farmers or managers was carried out, when they went to pick up the FMD vaccine to the offices of the Ministry of Agriculture and Fisheries, in February 2012; instance where by law all producers should vaccinate their cattle. The assessment included questions about the management practices of the animals at farms, characteristics of the facilities manage animals and staff training in animal welfare issues. The results showed that 97% of the farmers stated they have full and adequate facilities to handle cattle and 62% have loading facilities, of which half include shaded areas. Of all surveyed 75% say to use flags to move the animals; however, 66% say to use horses and 47% not specially trained dogs. Furthermore, 53% say not to practice dehorning in calves, they breed animals without horns. A total of 90% of farmers held they perform castration and of these 51% do it at very young age. The rest in older ages and only 4% use anaesthesia and/or analgesia. 90% of the surveyed farmers claimed to have heard about animal welfare and 42% attended update courses on the subject. However, at employee level, only 14% have attended a training course. Related to a traditional practice called 'yerra' in which all practices like castration, dehorning and hot iron branding are performed on the same day to animals of six months or more age, 76% said they no longer allow this practice in their farm. About devices to move animals, it is remarkable the widespread use of flags denoting a significant positive change in the perception of animal welfare by producers. This may be due to the joint efforts of institutions and more than 250 courses offered by the Continuing Education Program at the Veterinary School in collaboration with the OIE Collaborating Centre for Animal Welfare whose pro tempore headquarter is currently in Uruguay. It would be interesting to extend this survey to the whole country. It is clear that big improvements on animal welfare issues during the last few years have taken place in the country, but we should continue working on training and education mostly to rural persons to fully achieve the objectives proposed by the OIE standards.

Poster 6

An ethical framework applied to conducting research in working equids

C.E. Reix[1], P. Compston[1], M.M. Upjohn[1] and H.R. Whay[2]
[1]The Brooke, 41-45 Blackfriars Road, London SE1 8NZ, United Kingdom, [2]University of Bristol, Langford, Bristol BS40 5DU, United Kingdom; christine.reix@gmail.com

Working equids live predominantly in low-income countries. Conducting research into the welfare of these animals requires sensitive consideration of their working conditions and owners' livelihoods. Working equid owners are generally poor and, especially donkey owners, of low social status. Ethics is concerned with morality, beliefs, virtues, norms and what is right. Ethical frameworks, produced for use in biomedical science, can be applied to working equids, focusing on 4 principles: nonmaleficence, beneficence, autonomy and justice. This paper considers factors relating to research in working equids using the 4 principles ethical framework. (1) Nonmaleficence: Working equids are almost universally reported to be experiencing pain. Research methods may include physical examination of animals e.g. lameness assessment, involving exercise and limb manipulation, causing extra pain or distress. To minimise harm, power calculation should be performed to ensure only the number of equids needed for a study are included and considerate, minimally-invasive protocols employed. (2) Beneficence: Dehydration and heat stress are prevalent in working horses, water should be offered before and after examinations and shade provided, if possible. Research may include assessment but not treatment, therefore owners and animals should be linked to veterinary facilities for treatment if required. (3) Autonomy: Owners may be illiterate, or living in a hierarchical society and could be exploited. Owners should give informed consent (verbal and witnessed), decide whether to participate, be involved in the research process, including intervention planning. Results should be communicated back in an accessible and timely fashion. Owners may not want to take time out of their working day to participate in research as this leads to loss of income, for no immediate gain and an unknown future gain. Where possible animals should be given the chance to display natural behaviours: time to explore the examination area, roll on the ground when unhitched if desired; behaviours they are unable to exhibit in their daily work. (4) Justice: All animals and humans involved in the research study should be treated fairly and equally. Owner expectation should be managed, only offering that which is part of the study and possible outcomes explained. Care should be taken when advising owners, e.g. certain medicines or treatment protocols may be advised, but if owners cannot access, or afford them, then they cannot be implemented. In conclusion, using the 4 principles framework, specific ethical issues concerning research in working equids can be recognised, minimised and benefits maximised.

Poster 7

Assessment of welfare and animal sensitivity by actors of the community of the EISMV of Dakar

K.J.F. Adje[1], P.S. Kone[2], S.N. Bakou[3] and L. Mounier[4]
[1]Institut Pasteur de Côte d'Ivoire, Environnement et santé, 08 BP 1883 Abidjan 08, 225, Cote d Ivoire, [2]Ecole Inter-Etats des sciences et de Médecine Vétérinaire de Dakar, Sénégal, Santé et Environnent, BP 5077, Dakar 221, Senegal, [3]Ecole Inter-Etats des sciences et de Médecine Vétérinaire de Dakar, Sénégal, Santé et Production animale, BP 5077, Dakar 221, Senegal, [4]INRA, UR1213 Herbivores, 63122, Saint-Génès Champanelle, France; akjf17@yahoo.fr

The assessment of the sensitivity and the welfare of the animals are well treated. However, in African veterinary schools no study to our knowledge on the assessment of animal pain not was conducted. The objective of the study was to identify how the EISMV of Dakar actors perceive the sensitivity of animals and appreciate the importance of their well-being. The study was conducted at the school inter-States of Sciences and medicine veterinarians Dakar (EISMV). It is located at the University Cheikh Anta Diop of Dakar-Senegal. This work is a cross-sectional descriptive study. It took place from January to May 2011. Sampling was proportional random type. The opinion of respondents for each statement has been determined using the Likert scale. The data entry was made with the Epidata © 3.1 software. The χ^2 test and was used between the variables of interest at the 95% significance level. Statistical analysis software has been Rcommander © [version 2.12.0]. In total, 187 people were surveyed. More than 87% of people had heard once of animal welfare. Girls accounted for one third of the sample, students have formed 82%. About 70% of the people targeted were agreed that the welfare of the animals is a paramount issue. However, the opinion of respondents, based on the animal species being assessed was nuanced. More than 60% of respondents have been in disagreement with the statement that the hens, carnivores, pigs and ruminants do not feel pain as human. The perception of the animal pain sensitivity has not varied significantly according to sex, age, nationality, in the category of belonging, or the rank of the teacher (P>0.05). There is a difference in appreciation of the importance of animal welfare between the veterinary school of Dakar and the veterinary schools of developed countries. In the case of the assessment of animal pain, similarities exist

Poster 8

Dairy professionals' perspectives on assessing cow welfare on Canadian farms: a Delphi approach

C.G.R. Nash[1], T.J. Devries[2], J.B. Coe[1], E. Vasseur[3], D. Pellerin[4], D.F. Kelton[1] and D.B. Haley[1]
[1]University of Guelph, Population Medicine, Guelph ON, N1G 2W1, Canada, [2]University of Guelph, Kemptville Campus, Kemptville, Ontario, K0G 1J0, Canada, [3]University of Guelph, Alfred Campus, Alfred, ON, K0B 1A0, Canada, [4]Université Laval, Québec, QC, G1V 0A6, Canada; nashc@uoguelph.ca

More importance is currently being placed on animal welfare within the Canadian dairy industry than ever before, both in research and in practice. In support of this importance, the Canadian Code of Practice for Dairy Cattle has recently been revised (2009). Despite these advancements, it remains unclear how industry professionals are utilizing this knowledge and applying it on-farm. This study investigated how key expert groups associated with the Canadian dairy industry perceive cow welfare and identified areas of agreement and disagreement between them in relation to defining and assessing dairy-cow welfare. To accomplish these objectives a Delphi methodology was used: an iterative online survey process. The results from the first round of surveys are presented in this abstract. The survey was distributed to Canadian dairy professionals through e-mail lists and advertised in industry publications. The survey was available from September 1 to November 15 2013 in French and English. It included open- and close-ended questions relating to respondent demographics (n=15), use of welfare measures (n=4) and definition of dairy-cow welfare (n=2). A total of 113 surveys were completed and 7 professional groups were identified: 47 surveys were completed by veterinarians, 30 by extension workers, 28 by dairy producers, 18 by researchers, 8 by government workers and two by other groups. Seventy percent of participants were male. Mean years of dairy experience among participants was 20.6 years (range 2-59 years). Of these respondents, 99% specified they would use animal-based measure to assess dairy cow welfare, 86% would use resource-based measures, 79% would use management-based measures and 71% would use all three. When asked to define dairy-cow welfare, 98% of respondents included aspects of health, 89% included aspects of natural living, 87% included aspects of production, 86% include aspects of affective states and 78% included all four. There were no significant differences identified between professional groups on the selection of welfare measures or aspects identified in their welfare definition using Fisher's Exact Test. Findings demonstrate Canadian dairy professionals include the highly-debated aspects of production and affective state in their definition of dairy-cow welfare, with emphasis being place on animal-based measures to assess it. The results from this study show that Canadian dairy professionals have a cohesive view for defining and assessing dairy cow welfare.

Indicators for a result-oriented approach for animal welfare policies and organic farming

J. Brinkmann[1], S. March[1], C. Renziehausen[2] and A. Bergschmidt[2]
[1]*Thünen-Institute of Organic Farming, Trenthorst 32, 23847 Westerau, Germany,* [2]*Thünen-Institute of Farm Economics, Bundesallee 50, 38116 Braunschweig, Germany; jan.brinkmann@ti.bund.de*

Regulations in organic farming and in animal welfare policies are action-oriented: they refer to resources such as space allowance and management, i.e. access to pasture. With this approach, the prerequisites for the exercise of normal behaviour can be created; but direct impacts on animal health are largely left out. The aim of our project is to develop a concept for a result-oriented approach to improve animal welfare. One of the challenges is the selection of suitable indicators. Exemplarily for dairy, the presentation will suggest a set of indicators, approved by scientists and practitioners, which address major welfare issues in dairy farming. Although the Welfare Quality® protocols are considered to provide a valid and comprehensive assessment of animal welfare, their utilisation as a standard tool in organic farming or agricultural policy does not seem to be a realistic option. The time and skills required are major impediments for its widespread use. A full welfare assessment thus being out of reach, our proposal is to focus on the most important animal welfare problems. To select indicators which address major welfare-problems in dairy farming, we carried out a two-phased sampling process: (1) First, scientists from Germany, Switzerland and Austria (n=20) were asked in a written survey to select the most appropriate from a list of 82 indicators derived from literature. The questionnaire was designed as a Delphi survey to reduce heterogeneity among researchers and resulted in a list of 23 indicators. (2) Then a practitioner-workshop with farmers, lobby groups, administration and control took place (n=20) to address practicability issues. With great match, a list of 10 indicators was adopted. The following indicators were identified as adequate for providing robust evidence for improvements in animal welfare of dairy cattle to justify support payments (percentage cows beyond threshold on herd level). The Λ mark the approval of ⅔ of one group, the O approval of less than ⅔ but more than ½ of the group. Scientists | Practitioners Λ | Λ Somatic cell-count >400,000 ml -1 Λ | Λ Cleanliness Λ | Λ Fat-protein quotient >1.5 Λ | Λ Body condition Λ | Λ Lameness Λ | Λ Carpus- and tarsus alterations Λ | Λ Integument alterations O | Λ Lying behaviour: cow comfort index Λ | O Mortality of calves and cows Outlook The indicators are now subjected to a practical test on 150 dairy farms, in which the full Welfare Quality® protocol is carried out for validation and comparison.

Poster 10

A multicriteria decision analysis model for the selection of welfare strategies in dairy goat farms

A. Vieira[1], R. Amaral[2], G. Stilwell[1], T. Nunes[1], I. Ajuda[1] and M. Oliveira[3]
[1]CIISA-FMV, U. Lisboa, Avenida da Universidade Técnica, 1300-477 Lisboa, Portugal, [2]CEHIDRO-IST, U. Lisboa, Av. Rovisco Pais, 1049-001 Lisboa, Portugal, [3]CEG-IST, U. Lisboa, Av. Rovisco Pais, 1049-001 Lisboa, Portugal; ana.lopesvieira@gmail.com

When considering strategies to improve on-farm animal welfare, managers need to consider several aspects that represent the views of distinct stakeholders. Hence, there is a need for multi-criteria decision aiding methods to account for all of these perspectives. In this study we present a multi-criteria decision analysis model which was developed to assist the selection of welfare improving strategies in a large dairy goat farm. The model-building process followed a socio-technical approach, which combined social aspects of decision conferencing with technical components of building a multi-criteria additive evaluation model. This additive evaluation model was built with and to represent the views of the farm manager, in a set of decision conferences. In structuring the model, the first activity was the development of a cognitive map to identify and structure key concerns relevant for the selection of strategies. Six key concerns were identified: animal welfare, milk quality, farm image, profit, impact of the strategies on farm routines, and present results in a short-medium term. Key concerns were made operational by means of descriptors of performance, in which the neutral level was defined as the status quo. For example, performance on milk quality was depicted by the somatic cell count (a quantitative descriptor); and for profit, an economic model of impact for measuring the performance of different strategies was built. The Measuring Attractiveness by a Categorical Based Evaluation Technique (MACBETH) and the M-MACBETH decision support system assisted in the construction of the (multi-criteria) evaluation model. The distinctive characteristic of MACBETH is that it requires qualitative judgments about differences in attractiveness between performance levels of the descriptors, and later among hypothetical strategies, to build an evaluation model with value scores and weighting coefficients. Once value scores and weights were validated by the decision-maker, MACBETH was also used to perform sensitivity and robustness analyses, and to comparatively analyse the performance and value of the different animal welfare strategies when individually or simultaneously taken. As a result from developing the model, the decision-maker could conclude that the most attractive strategy was one that combined two different measures aiming at the welfare conditions of lame and overweight goats: 'hoof trimming and diet change'. The developed constructive decision-aid approach provided a transparent and effective way of comparing animal welfare strategies.

The use of focus groups to consult equine stakeholders about welfare assessment

S. Horseman[1], J. Hockenhull[1], S. Mullan[1], H. Buller[2], A. Barr[1], T.G. Knowles[1] and H.R. Whay[1]
[1]University of Bristol, School of Veterinary Sciences, BS40 5DU, United Kingdom, [2]University of Exeter, Geography, College of Life and Environmental Science, EX4 4RJ, United Kingdom; sue.horseman@bristol.ac.uk

In recent years the value of stakeholder consultation in the development of welfare assessment protocols has been recognised. Consultation allows stakeholder views to be combined with scientific observation to maximise the relevance, reliability, validity, brevity and feasibility of the assessment protocols developed. As part of work to develop a welfare assessment protocol for use in horses, focus groups were conducted with six groups of equine stakeholders. The purpose was to gain insight on stakeholder views as to what should be included in a welfare assessment and how it could or should be executed. The focus groups were designed to facilitate discussion and allow both agreement and disagreement to be aired. Separate focus groups were run with leisure horse owners, grooms, professional riders, equine vets, equine welfare charity workers and equine welfare scientists. As a tool for allowing stakeholders to identify and discuss welfare needs the focus groups proved very successful. Stakeholders generally felt confident identifying welfare needs although different groups placed emphasis in different areas. The focus groups were less useful in facilitating a discussion about welfare assessment. Many stakeholders found it hard to understand the value of assessing welfare beyond using it as a way to identify and correct poor welfare in an immediate context as opposed to a tool for identifying wider, ongoing welfare concerns. They believed that welfare assessment would only be useful if it was conducted in places where they assumed welfare would be poor and questioned whether a researcher would get access to such places. Many groups also found it hard to identify objective welfare assessment methods and saw the process of welfare assessment more as a process of gathering a subjective impression rather than as a systematic process. The focus groups highlighted the need to promote more objective welfare assessment methods to equine stakeholders, who may have difficulty understanding the range of contexts in which welfare assessment can be used. This may facilitate a broadening understanding of the value of welfare assessment beyond supporting pre-conceived ideas of what good and poor welfare is and where and when it may occur. Acknowledgements: This work was funded by World Horse Welfare and the authors would like to thank the members of the equine industry who participated in the study.

Poster 12

Survey of stakeholder opinion on dairy cow welfare assessment schemes: measuring and reporting

M.J. Haskell and A.C. Barrier
SRUC, Animal and Veterinary Sciences, West Mains Road, EH9 3JG, Edinburgh, United Kingdom; marie.haskell@sruc.ac.uk

Understanding how all stakeholders in the industry view the whole process of welfare assessment is critical to having assessment schemes accepted by the industry and the public. For a welfare assessment scheme to be accepted by all stakeholders, it must fulfil a number of criteria. It must measure important aspects of welfare, and measure those aspects adequately and reliably. The level of involvement of the farmer during assessments and the way in which the results are presented to the farmer, and are used by the scheme owner, are also important. Therefore, the aim of this study was to collate stakeholder views on how dairy cow welfare assessment schemes can be delivered to maximise acceptance. A review of welfare assessment schemes and measures was used to devise an on-line questionnaire containing 49 questions. The questions asked for the respondent's opinion on: how the assessment procedure should be carried out, who should pay for assessments, a rating of the importance of different welfare criteria (e.g. mobility, disease status), the reliability and practicality of current welfare measures within the criteria and how the results should be used and presented. The groups of stakeholders invited to take part were: representatives from the dairy industry/trade organisations, dairy research scientists, dairy advisors, dairy veterinarians, dairy farmer representatives, consumer and animal welfare charity organisations, milk buyers and retailers. There were 108 responses from the 175 invitations to answer the questionnaire. There was little consensus on who should pay for assessments, but milk retailers, milk purchasers and consumers were most favoured (67, 62 and 51% respectively). Mobility, presence of disease and presence of injuries were scored as the most critical aspects of welfare to assess (rated as extremely important by 86, 79 and 66% of respondents respectively). Cleanliness and 'living of a natural life' were the least important (23 and 12%). When a non-compliance against a scheme standards is found, most respondents favoured actions allowing the farmer to discuss it with the assessor (73%), or the farmer being given time to perform any required action (65%). There was less support for penalising farmers (35%) and little support for the farmer using the information as they see fit without any further action from external sources (2%). There was a high response rate to the questionnaire indicating interest across the dairy industry. There is broad agreement on what aspects of welfare to measure across all stakeholder groups. It was clear that the stakeholders view welfare assessment protocols as tools for use by the farmer on his/her own farm as well as a tool for certification bodies or other assessors.

Managing consultations advisory board: pilot coordinated European animal welfare nnetwork EUWelNet

D.G. Pritchard[1], H.J. Blokhuis[2], J.C. Jinman[1] and I. Veissier[3]
[1]Veterinary Consultancy Services Limited, Argyll House, SW18 1EP London, United Kingdom, [2]Swedish University of Agricultural Sciences, Arrheniusplan 3, Uppsala, Sweden, [3]INRA, Clermont, Saint Genès Champanelle, France; davidgeorgepritchard@gmail.com

The EUWelNet project evaluated the feasibility and usefulness of a network to assist competent authorities and stakeholders in implementing EU legislation on animal welfare. See www. euwelnet.eu. The Advisory Board (AB) was made up of representatives of 27 Member States, Croatia, Norway, and Switzerland, EU institutions (DGSANCO, EFSA, FVO), international organisations (OIE, FAO, EUROFAWC), European animal and meat industries (Copa-Cogeca, EFFAB, IFAH, UECBV), veterinary and welfare science (FVE, IASE, ISAH), welfare education (EVVPH, ECAWA,), animal welfare organisations (CIWF, Eurogroup, VierPfoten, WSPA), European Animal Welfare Platform (FAI).This broad-based AB reflected the project's emphasis on pig, broiler and killing legislation. The wide stakeholder input via members' networks helped the AB to ensure the transparency, relevance, validity, reach and quality of the work. Close involvement in the project provided the AB with a basis to give firm and independent input on the project's recommendations on a future animal welfare network. The first AB meeting had a project update and agreed the advice required by the project team using open discussions in small groups with leadership and challenge of outcomes at plenary sessions to seek agreement. Conflicts were minimised by a code of practice for members and by members acting as 'critical constructive friends'. Consultations of the AB were managed by an independent AB chairman with task leaders and was tailored to the specific project tasks. AB members facilitated access to key sources, provided detailed technical information, completed questionnaires, attended workshops and evaluated best practices, e-learning and other educational material. They provided advice on relevant international organisations and networks. The results were remarkable for the enthusiasm generated, information exchanged and the improvements in the project outputs. The second AB meeting reviewed proposals for structures of a future European network, its operation, feasibility, scope and role, efficient monitoring of implementation of the EU legislation and valid data collection. The AB concluded that technical excellence and independence from private individual interest groups could be sustained by a future network. The AB was a useful friendly platform which improved both the cooperation of the project team with stakeholders and the agreement on the optimal structure of a future Network of European welfare reference centres which balances the needs of regions and species expertise.

Poster 14

Qualitative behavioural assessment of variation across time in the demeanour of individual hill sheep

S.E. Richmond, C.M. Dwyer and F. Wemelsfelder
SRUC, Roslin Institute Building, EH25 9RG Easter Bush, United Kingdom;
susan.richmond@sruc.ac.uk

Qualitative behavioural assessment (QBA) is a whole-animal approach, integrating perceived details of animals' expressive demeanour, using terms such as tense, anxious, or relaxed. Accommodating fluctuations in animal welfare over time is an important challenge for welfare assessment protocols. The aim of this study was to investigate the efficacy of QBA in detecting variation in the demeanour of individual hill sheep across a pre-lambing/post-weaning time period. A list of 21 QBA terms for sheep was developed by an expert focus group. Significant agreement was found between 3 observers in applying this list to 48 Scottish Blackface ewes on a Scottish hill farm. One of these observers scored the same 48 ewes over 13 visits from March to August 2013. Visits 1-4 took place pre-lambing, visits 5-11 post-lambing, and visits 12-13 post-weaning. QBA scores for the 13 visits were analysed together using Principal Component Analysis (correlation matrix, no rotation). The effects of visit number, time of day (am/pm), and ambient temperature on PC1 scores were investigated using random-coefficient modelling. In addition, the effects of pre/post-lambing and post-weaning time periods on the first 4 PC scores were investigated by repeated measurements ANOVA. PCA distinguished 4 meaningful dimensions of sheep expression: PC1 (23%) 'content/bright-apathetic/subdued'; PC2 (17%) 'relaxed/calm-tense/agitated; PC3 (12%) 'wary/fearful-listless/apathetic', and PC4 (9%) 'wary/fearful-aggressive/defensive'. There was a significant effect of visit number on PC1 ($P<0.001$), showing sheep to gradually become more 'content/bright' with the advance of summer. Within this shift, when corrected for effect of visit number, ambient temperature showed a trend towards a significant effect on PC1 ($P<0.084$), indicating that as temperatures rose ewes became more content/bright, yet above 19 °C they grew more apathetic/subdued. There was no statistical evidence for a time of day effect on PC1. There were significant effects of lambing/weaning time periods on all 4 PCs ($P<0.001$). The effect on PC1 supported the gradual improvement of mood as time advanced, with mood lowest pre-lambing and highest post-weaning. However, in addition, effects on PCs 2, 3 and 4 showed that after lambing, sheep became significantly more tense/agitated, listless/apathetic, and wary/fearful respectively, a finding in accord with the reduced health of the flock at this time. There was no evidence for consistent individual differences in sheep scores on any PC over these periods. These results suggest QBA is capable of addressing different levels of variation affecting the mood of sheep over time, and may be useful in fine-tuning and standardising welfare assessment protocols.

The relationship between awake+sleep states and autonomic nervous balance in cattle

K. Yayou[1], K. Watanabe[2], M. Ishida[3], E. Kasuya[1], M. Sutoh[4] and S. Ito[2]
[1]National Institute of Agrobiological Sciences, Division of Animal Sciences, Tsukuba, Ibaraki, 305-8602, Japan, [2]Tokai University, School of Agriculture, Aso-gun, Kumamoto, 869-1404, Japan, [3]Institute of Livestock and Grassland Science, Animal Waste Management and Environment Research Division, Tsukuba, Ibaraki, 305-0901, Japan, [4]Institute of Livestock and Grassland Science, Animal Physiology and Nutrition Research Division, Tsukuba, Ibaraki, 305-0901, Japan; ken318@affrc.go.jp

To contribute to make simple identification method for awake+sleep states in cattle, we examined the relationship of autonomic nervous balance with awake+sleep states categorized by behavioural observations. Six Holstein steers aged 8-10 months were used. The awake+sleep states were categorized by the combination of three behavioural measures: body posture, neck position, and eye status, into 'awake', 'drowsy', 'non-rapid-eye-movement sleep (NREM)', 'rapid-eye-movement sleep (REM)', and 'sleep state which could not be categorized into neither NREM nor REM (UC)'. Power spectral analyses of R-R interval variability of electrocardiogram obtained by radio telemetry were performed to calculate the low frequency (LF, 0.04-0.1 Hz) and high frequency (HF, 0.1-0.8 Hz) powers. There were differences in LF/HF (P<0.1), which represents sympathetic activity, and HF in normalized unit (HF/[LF + HF] × 100) (P<0.05), which represents parasympathetic activity, among awake+sleep states. The autonomic nervous balance shifted toward parasympathetic dominant during NREM and UC and toward sympathetic dominant during awake and REM. Our findings about awake+sleep states-dependent variation in autonomic nervous balance are in line with previous evidences in humans. Since the autonomic nervous balance was similar between NREM and UC, UC could be categorized into NREM. Though these results should be validated by further study utilizing electroencephalogram, combined use of behavioural observation and autonomic nervous balance could be a useful identification method for awake+sleep states in cattle.

Poster 16

Adrenal responce to stress in rabbits

M. Dyavolova, P. Moneva, I. Yanchev and D. Gudev
Institute of Animal Scince, Spirka Pochivka, 2232, Kostinbrod, Bulgaria; m.dyavolova@gmail.com

The purpose of this study was to investigate adrenal response to stress in 4 months old rabbits. Ten New Zealand male rabbits, accommodated individually in wire-floor cages were exposed to various stressors: dog barking, bringing a dog into the rabbit house, forced running for 10 min., injection of adrenocorticotropic (ACTH) hormone. The rabbits were reared in enclosed building under naturally fluctuating temperatures. Plasma cortisol, corticosterone (measured by Rabbit cortisol Elisa kits), and heterophil/lymphocyte (H/L) ratio were determined before and following exposure to the corresponding stressors. Plasma cortisol and corticosterone levels in the rabbits, unlike most other mammals, were not increased after running, administration of ACTH (100 µg/rabbit), and presence of a dog. On the contrary, ACTH injection caused decline in plasma cortisol to levels that were under the sensitivity of the method. Dog barking was the only stressor that caused significant increase in plasma corticosterone at 30 and 60 min, following the end of stressor treatment, but unexpectedly it decreased H/L ratio. Surprisingly, H/L ratio increased significantly ($P<0.01$) at 1 and 2 h following ACTH injection. Our data suggest that rabbits don't respond to stress by the classical increase of glucocorticoid levels. The observed increase of H/L ratio against the background of declining cortisol and corticosterone levels following ACTH injection suggests that ACTH itself is implicated in the modulation of H/L ratio. The results cast doubt on the reliability of cortisol and H/L ratio as indirect markers of stress in rabbits.

On farm welfare assessment in dairy goat farms

S. Waiblinger, C. Graml, E. Nordmann, D. Mersmann and C. Schmied-Wagner
University of Veterinary Medicine Vienna, Department of Farm Animals and Veterinary Public Health, Institute of Animal Husbandry and Animal Welfare, Veterinärplatz 1, 1210 Vienna, Austria; susanne.waiblinger@vetmeduni.ac.at

Although still low compared to dairy cows the number of dairy goats has increased over the last decade in several European countries due to increasing demand for goat milk products and welfare assessment protocols may help in identifying problems, improving the situation and enable labelling. In the course of a specific welfare question – comparing herds consisting only of hornless goats with herds with mixed horn status – we used a comprehensive protocol to assess the welfare of goats and influencing factors on 45 dairy goat farms (herd size 78-518 lactating goats). The aim of this presentation is to give some results regarding welfare problems on the farms investigated, an overview on measures and to discuss their completeness and usefulness for an overall welfare assessment. For this we also compare the measures used in our protocol to the Welfare Quality® protocol for dairy cows. Most of the 12 welfare criteria making up the four welfare principles in the Welfare Quality® protocols were covered by animal based measures in the goats, and similar or identical measures were used except for the principle good housing, due to quite different housing systems and different welfare problems and needs in cows and goats (e.g. no tethering in dairy goats, provision of heightened lying platforms as special need for goats). Expression of other behaviours was adapted to goat needs – possibilities for climbing and lying heightened besides access to pasture. The measures used revealed large between-farm variability. For example, the percentage of animals being too thin (BCS<2.5) ranged from 0 to 64%, the number of altered lymphnodes from 0-82%, social interactions with body contact ranged from 0.1 to 1.2 interactions / goat / 10 min, and the average distance goats kept to an unfamiliar experimenter was between 0 and 4 m. When benchmarking farms into thirds, no farm was found in the better third of farms in all criteria, but at least one of the criteria was in the lowest third in all farms indicating farm specific problems. The measures used give a quite complete assessment of goat welfare and were sensible to identify problems and differentiate farms. However the measures are relatively time – consuming and assessing them in an order to avoid carry–over effects even worsen this situation. Further development should focus on optimizing these aspects, testing reliability of the whole protocol and, if necessary for some purposes, model an overall welfare score. We acknowledge funding by BMLFUW and BMG, Project-Nr.100191.

Poster 18

Recording organisations and collection of animal-based welfare indicators

P.L. Gastinel[1], E. Berry[2], C. Egger-Danner[3], L. Mirabito[4] and K.F. Stock[5]
[1]France Génétique Elevage, 149 rue de Bercy, 75012 Paris, France, [2]DairyCo, Stoneleigh Park, Kenilworth, Warwickshire, CV8 2TL, United Kingdom, [3]ZuchtData EDV-Dienstleistungen GmbH, Dresdner Straße 89/19, 1200 Wien, Austria, [4]Institut de l'Elevage, 149 rue de Bercy, 75012 PARIS, France, [5]VIT – Vereinigte Informationssysteme Tierhaltung w.V., Heideweg 1, 27283 Verden, Germany; pierre-louis.gastinel@laposte.net

The recording organisations around the world collect data on individual animals in several hundred thousands of farms, with the aim of improved farm management and genetic selection. These data cover traits with a long recording history like milk yield and composition, beef production, conformation traits and increasing numbers of functional traits. Functionality-related information on reproduction, health and other novel traits are important for sustainable livestock management and to satisfy social expectations, and are linked to some welfare traits. The International Committee of Animal Recording (ICAR) has 105 members who are involved in data collection and genetic evaluation of 30 million animals from 53 countries, in 5 continents. The object of ICAR, an international non-profit body, is to promote the development of performance recording for farm animals and their use in genetic evaluation and farm management. This object is achieved through establishing definitions and standards for measuring characteristics of animals. Specific guidelines for identifying animals, recording their parentage and their performance, evaluating their genetics are elaborated, and findings are published. OIE, intergovernmental organisation, established «General Principles for the Welfare of Animals in Livestock Production Systems» and, for different species, 'Criteria or Measurables for the Welfare'. Our paper proposes an crossed analysis between the general principles of OIE, and its criteria, on the one side, and the practices of the recording organisations based on ICAR guidelines, on the other. We identified: convergence points using data already collected in the herds, sent in the recording and genetic databases, potentially usable to evaluate the trend of animal welfare; complementary points, where ICAR guidelines give a description, with a harmonization of the data allowing exchange and building-up of central databases with related infrastructures; and gaps, where the ICAR Guidelines may be complemented by new indicators more appropriate to monitor animal welfare that can be generated by recording organisations. Finally the importance of interdisciplinary exchange and collaboration between scientists, governmental bodies, farmers, breeding and recording organisations in this field is reiterared.

Assessment score of mink in the nursing period decreases with date of assessment

B.I.F. Henriksen and S.H. Møller

Aarhus University, Department of Animal Science, Blichers Allé 20, Postboks 50, 8830 Tjele, Denmark; britt.henriksen@agrsci.dk

The objective of the present study was to test the hypothesis that the score of the four welfare measures, 'Body condition', 'Dirty nests', 'Injuries' and 'Diarrhoea', changes significantly with the date of assessment within the data collection period from parturition to weaning, influencing the scores of WelFur at criteria level. We further expect, however, that the number and magnitude of changes will not be enough to change the welfare score at the principal level or the overall classification of mink welfare according to the WelFur-Mink protocol. Data from a representative sample of 120 dams and litters from each of four farms was collected three to four times in the period from parturition to weaning according to the WelFur-Mink protocol. The data was collected by the same, experienced external personnel at all assessments. WelFur-scores between 0 (worst) and 100 (best) were calculated, aggregated, and compared at criteria and principal level. The frequency of very thin animals, dirty nests and injuries did significantly increase with the date of assessment within the data collection period, while there were no significant changes in the frequency of diarrhoea with the date of assessment. The WelFur protocol is based on aggregation across three periods while our calculations are for one period only. For the two other periods we therefore used the average values from WelFur-assessment on nine mink farms to simulate how much the measurements that changes with age contribute to the annual score per principle and to the overall classification. The score for the criteria 'Absence of prolonged hunger' dropped significantly from 80 to around 40 after about five weeks of lactation, affecting the principal score 'Good feeding'. The score for the criteria 'Absence of injuries' was significantly lower after about six weeks of lactation, but did not influence the principal score 'Good health' significantly. The overall classification changed from 'Best current practice' to 'Good current practice' after about six weeks of lactation. In conclusion, the overall WelFur classification of farms changes with time of assessment in the nursing period. This is mainly due to an increase in very thin animals at the end of the lactation period. A less time-dependant WelFur-assessment can be achieved by reducing the time window for assessment to either before or after five weeks pp. when the score for 'Absence of prolonged hunger' changes or by development of a valid correction factor for Body condition score. Shortening the assessment period will hamper the use of WelFur so the best option will be to develop a correction factor in order to maintain this important period in the WelFur-assessment of mink farms.

Poster 20

Is the duration of the lying down and getting up movements modified by the softness of the floor?

L. Mounier, M. Touati, E. Delval and I. Veissier

Inra / VetAgro Sup, 1 avenue Bourgelat, 69280 Marcy l'étoile, France; luc.mounier@vetagro-sup.fr

In dairy cattle, low resting time may be associated with discomfort and can increase health problems. Most of the research has shown that housing facilities can reduce cow lying time. The total lying time, the frequency and duration of lying bouts are the most common indicators to assess the lying behaviour. Nevertheless, even if automated methods of recording may be useful, f these indicators are not feasible on-farm. Recently, the duration of the lying down movement has been found to be reliable for on-farm assessments and to allow distinction between different housing systems. When cows lie down and get up, they place much of their weight on their knees which could make the movements painful and therefore may modify the duration of these movements. However, to our knowledge, no studies are available comparing the effects of flooring softness on the duration of lying down and getting up movements. The purpose of this study was to examine the effect of floor softness on the duration of the getting up and lying down movements of lactating dairy cows. Sixteen cows were filmed during two weeks from 06:00 to 22:00 h. They were allowed to choose between three different floors on cubicles: 6 cubicles with very soft flooring (VS) (mattress covered with 2 kg sawdust each day), 5 with soft flooring (S) (mats covered with 0.5 kg sawdust) and 5 concrete covered with 1 kg sawdust (C). BCS, locomotion score, height and length of the cows were recorded. We used a mixed model at animal level to analyse the links between the getting up and lying down movements and the explanatory variables (softness and individual factors of the cows). The C cubicles were occupied 4.60% of the time, the S cubicles 35.42% and the VS cubicles 59.98%. Eight cows were never lying on C cubicles and two cows were lying only on VS cubicles. The softness had a significant effect on the duration of the lying down movements (P=0.005) with a duration shorter when the cows lied on the S cubicles (P=0.0009; 4.67 s in S vs 4.87 s in VS cubicles). The duration of the lying down movements varied largely within cows, for instance from 3.4 to 13.52 s to lie down in VS cubicles. The softness had no significant effect on the duration of the getting up movement; this duration had a tendency to be longer when the locomotion score was higher. The comfort of each type of flooring was confirmed by the occupational time of the cubicles. The duration of the lying down may be modified with the softness of the flooring. However, the variations between softness classes were limited and the high variability within cows make this indicator difficult to use for on-farm assessment of the softness of the bedding area.

Genetic correlations of peri-natal behaviour with lamb survival in South African Merinos

S.W.P. Cloete[1,2], J.J.E. Cloete[3] and A.J. Scholtz[1]
[1]Western Cape Department of Agriculture, Directorate Animal Sciences: Elsenburg, Private Bag X1, 7607 Elsenburg, South Africa, [2]University of Stellenbosch, Department of Animal Sciences, Private Bag X1, 7602 Matieland, South Africa, [3]Elsenburg Agricultural Training Institute, Private Bag X1, 7607 Elsenburg, South Africa; schalkc@elsenburg.com

Lamb survival is increasingly regarded as indicative of the welfare of sheep flocks. Data from a South African sheep flock suggested that selection for the composite trait, number of lambs weaned per lambing opportunity, also resulted in correlated responses in age specific lamb survival. Moreover, such selection also resulted in genetic changes in peri-natal behaviour that were regarded as conducive to lamb survival. However, genetic correlations of lamb survival with behaviour traits are scarce in the ovine literature. Monte-Carlo Markov Chain algorithms employing Gibbs sampling was used to estimate genetic correlations of lamb survival (assessed on the underlying liability scale) with recorded behaviour traits in a South African Merino flock that was divergently selected for the ability of ewes to rear multiples. Between 2,586 and 2,705 individual lamb survival records and between 1495 and 1621 individual lamb behaviour records were used for this purpose. Heritability estimates were 0.17 for the survival of birth, 0.14 for perinatal survival, 0.16 for survival from tail-docking to weaning, 0.10 for the duration of birth (the interval from the first sign of impending parturition to the expulsion of a specific lamb), 0.23 for the interval from the birth of a specific lamb to standing for >10 seconds, and 0.12 for the interval from standing to first sucking for >10 seconds. Corresponding maternal heritability estimates amounted to 0.21, 0.17, 0.17, 0.13, 0.09 and 0.07. Genetic and maternal genetic correlations of the interval traits with lamb survival at all stages were mostly negative (i.e. favourable), ranging from -0.32 between the interval between standing to suckling and survival from tail-docking to weaning to -0.58 between the duration of birth and the survival of birth. Corresponding maternal genetic correlations ranged from -0.03 to -0.63. The cooperation of the ewe with the first suckling attempts of the lamb was also favourable related to the interval from standing to suckling being -0.41 on the direct genetic level and -0.44 on the maternal genetic level. However, these correlations were associated with comparatively large standard errors and rarely significant. Although the absolute magnitude of the vast majority of these direct and maternal genetic correlations were favourable, further analyses, employing larger databases, are indicated to obtain more accurate estimates, thereby verifying the size and magnitude of the respective correlations.

Heart rate variability: its significance as a non-invasive stress parameter in dairy cattle welfare

L. Kovács[1,2], L. Kézér[1], V. Jurkovich[2], O. Szenci[2] and J. Tőzsér[1]
[1]Szent István University, Faculty Agricultural and Environmental Sciences, Páter K. u. 1., 2100 Gödöllő, Hungary, [2]Szent István University, Faculty of Veterinary Science, István utca 2, 1078 Budapest, Hungary; kovacs.levente@mkk.szie.hu

In the past, housing technologies and management practices for dairy cattle were mainly assessed by descriptive behavioural methods. Non-invasive techniques for cortisol assays also provide reliable results; however, sample collection is limited in loose housed cows. In the last decades – complementary to behavioural and physiological stress measures – heart rate (HR) and heart rate variability (HRV) measurements enable continuous data recording by monitoring the autonomic nervous system activity, that has recently gained considerable interest worldwide in farm animals. HR measurements have been used to evaluate stress in farm animals since the beginning of the 1970's. Nevertheless, according to recent studies in veterinary and behaviour-physiological sciences HRV proved to be more precise for studying the activity of the autonomic nervous system. In the last 15 years the use of HRV analysis for examining welfare has become prominent in livestock species. In dairy cattle, an increasing number of studies used parameters of HRV to indicate stress caused by diseases, routine management practices, milking and painful procedures in calf rearing. This work provides the significance of HR and HRV measurements in dairy cattle research by summarizing current knowledge and research results in this area. First, the biological background and the interrelation of the autonomic regulation of cardiovascular function, stress, HR and HRV are discussed. Equipment and methodological approaches developed to measure interbeat intervals (IBIs) and estimate HRV in dairy cattle are described. The methods of HRV analysis in time-, frequency- and non-linear domains are also explained emphasizing their physiological background. Even though HR and HRV measurements are not untouched fields in dairy cattle welfare studies, large areas are still left for future research, therefore, finally, the most important scientific results and potential possibilities for future research are also presented. In conclusion the strength of their sensitivity, HR and HRV measurements may improve our ability to assess and interpret autonomic nervous system responses to short-term and chronic stress and have significant potential to provide valuable insight also on pathological conditions of dairy cattle. Focusing on the aspects of stress that are most relevant to the animal's welfare may provide valuable information on how dairy cattle handling and housing can be improved in the near future.

How to assess sheep welfare on pasture?

M.M. Mialon[1,2], A. Brule[3], C. Beaume[4], A. Boissy[1,2], D. Gautier[4], D. Ribaud[3], L. Mounier[1,2] and X. Boivin[1,2]
[1]*VetAgro Sup, 1 Av Bourgelat, 69280 Marcy l'étoile, France,* [2]*INRA, UMRH1213, Theix, 63122 St Genes-Champanelle, France,* [3]*Institut de l'Elevage, Monvoisin BP 85225, 35652 Le Rheu, France,* [4]*CIIRPO, Ferme du Mourier, 87800 St Priest Ligoure, France; marie-madeleine.richard@clermont.inra.fr*

Animal-based protocols are still rare or under development in sheep and not yet validated in farming conditions for animals reared outdoor. The aim of this presentation is to summarise the development of a welfare assessment protocol for sheep during pasturing and to test elements of validity of this assessment. Several animal-based measures inspired from the 4 principles of Welfare Quality® were selected regarding their feasibility on farm and performed on Romane sheep. At an individual-level the body condition score (Feeding), animal dirtiness and wool humidity (Housing) and lameness, lesions, respiratory disorders, udder (Health) were assessed for their inter- and intra-observer reproducibility. For that, 30 randomly selected ewes per farm were tested on 10 farms. For the principle Appropriate Behaviour, a human approach test was performed on three 30-ewes groups per farm. Intra-class correlation or kappa coefficients were computed. Then, between-farms variability was estimated on 53 farms in summer time by variance analysis with a mixed model. Finally this protocol is presently applied in 8 farms to evaluate sheep from the same original flock split in two conditions in each farm (pasture, n=50 vs barn, n=50). Three assessments are performed at 0, 3 and 6 weeks after splitting. Inter-observer reproducibility was good for all measures (>0.66). Intra-observer reproducibility was generally good for individual measures (>0.65) but was poor (<0.2) for the human approach test. In summer time, most measures presented occurrence rates below 2% except lameness (5.6% of ewes) and nasal discharge (3.6%). The highest occurrence rate was noticed for animals' back dirtiness (26%). With this low level of occurrence, within-farm variability was higher than between-farms. By contrast, between-farms variability represented more than 60% of the total variability for the human approach test, suggesting a promising measure for discriminating farms. The data analysis comparing indoor and outdoor housing during winter time is still in progress and will be achieved at the congress date. In sheep, it is necessary that animal welfare is also ensured at pasture, in particular in winter time. According to our results, the recorded criteria could be used to assess welfare at pasture and reveal few problems occurring in summer time. The comparison of the data recorded in winter will provide information on potential differences in welfare between barn and pasture. Our work should help to evaluate the sustainability of wintering at pasture for sheep in France.

Test of procedures for observation of stereotypy in mink during winter

S.H. Møller, S.W. Hansen and B.I.F. Henriksen
Aarhus University, Department of Animal Science, Blichers Alle 20, 8830 Tjele, Denmark;
steenh.moller@agrsci.dk

In a welfare assessment protocol for mink (WelFur), stereotypy is one of three measurements under the criteria 'Expression of other behaviours' within the principle 'Appropriate behaviour'. The observation procedure was developed from procedures described ion scientific publications. The controlled experimental conditions, however, proved difficult to apply in practice. In the development of WelFur we tried to standardize the observation procedure by postponing the feeding and take the observations at the time the mink would normally have been fed. This turned out to be problematic for three reasons: Firstly, in some periods farmers feed two or more times a day. Secondly, not all farms fed the same time every day. Thirdly, if they did, they would often feed other mink on the farm during the observation of stereotypy, which seemed to affect the observations. In order to avoid these problems we moved the observations to one hour before the main feeding. We have therefore tested the effect of this change in observation in relation to feeding. Furthermore we tested how the stereotypy observation procedure could be simplified and applied under less standardised conditions without a significant loss of information. In order to reduce the time needed for taking the measurement an observation period of one minute was tested against the two minutes defined in WelFur. In order to simplify the conditions for observation, the effect of a feeding machine running within hearing distance of the mink during observation was tested. The procedures were tested in late February as described in the WelFur-mink protocol for the winter period. The preliminary analysis of data indicates that the stereotypy observations taken one hour before feeding did not differ from the observations taken when feeding was postponed. The revised procedure in the WelFur protocol does thus not affect the outcome of the stereotypy observation during winter. The sound of a feeding machine seemed to increase the number of active as well stereotyping mink observed. It was therefore correct to standardize the observation procedure to before feeding started and it is important to complete the observations before feeding begins. A significant number of mink was observed performing stereotypy in the second minute of the observation period only, causing the distribution of stereotyping and non-stereotyping mink to be significantly different between one and two minutes observations. The observation time can therefore not be reduced from two to one minute during the winter period without loss of information.

Transect walks: method sensitivity for on-farm welfare evaluation in turkeys

T.T. Negrao Watanabe[1], V. Ferrante[1], I. Estevez[2,3], C. Tremolada[1], L. Ferrari[1], S. Lolli[1] and R. Rizzi[1]
[1]Università degli Studi di Milano, Department of Veterinary Science and Public Health, via Celoria 10, 20133 Milan, Italy, [2]KERBASQUE, Basque Foundation for Science, Alameda Urquijo 36-5, 48011 Bilbao, Spain, [3]Neiker-Tecnalia Animal Production, 01080, Vitoria-Gasteiz, Spain; tatiane.negrao@unimi.it

There is increasing consumer demands for livestock and poultry products that meet the minimum expectations in terms of animal welfare during their production cycle. Additionally, a growing number of farmers are aware about full compliance with the vital animal welfare standards that, in addition, could play an important economic role in commercial intensive productions. Indeed, animal welfare assessment protocols have meaningful effects on providing the legal verification in order to promote and guarantee the on-farm safeguard animal standards. Transect walks method appeared to provide a practical approach to welfare assessment in broilers farms. Because of the generalities of meat poultry production this method could be consider a reasonable approach for turkey welfare evaluation in terms of time demands, within a reasonable costs, and it is not physically demanding. The aim of this study was to determine the sensitivity of this method in commercial turkey flocks (Animal Welfare Indicators project, FP7-KBBE-2010-4). In this study 10 commercial turkey female flocks (6 houses), ranging 6,000 to 7,200 birds and belonging to the same company were evaluated one week before slaughter. On turkey farms, walking through the house is a routine daily procedure to check the health status of the birds. Two previously trained assessors in performing the transect methodology and properly assessing the selected indicators, evaluated each paired house sequentially and independently within the same day by walking through predefined transect bands (1 to 4) in random order. The animal-based indicators considered were: immobile, limping, wounds, featherless. The statistical model used was GLM in the GENMOD procedure. The results showed that this welfare assessment approach highlights even small variation among houses for the considered variables. In fact there were significant difference across houses (P<0.008). Differences across observers were detected for wounds and fatherless (P<0.0001), results that may be due to the difficulties in assessing equally these parameters while walking. On the contrary lameness, probably the most important welfare problem in meat poultry, shows a good concordance between observer (P=0.361). These preliminary findings, suggest that this new approach has potential as a tool for on-farm welfare evaluation, which may be worthwhile to be further developed.

Study on gastric ulcers in heavy pigs receiving different environmental enrichment materials

E. Nannoni[1], M. Vitali[1], P. Bassi[1], L. Sardi[1], G. Militerno[1], S. Barbieri[2] and G. Martelli[1]
[1]University of Bologna, Department of Veterinary Medical Sciences, Via Tolara di Sopra 50, 40064 Ozzano Emilia (BO), Italy, [2]Università degli Studi di Milano, Department of Veterinary Science and Public Health, Via G. Celoria 10, 20133 Milano, Italy; giovanna.martelli@unibo.it

Given the negative effects that stress, among other factors, can exert on swine gastric mucosa integrity, the presence and the severity of gastric lesions were assessed post mortem on pigs that during their lives had received two different forms of environmental enrichment. To this aim 56 hybrid male pigs (initial average BW: 30 kg) were homogeneously allocated to 2 experimental groups: one half of the animals was provided with hanging chains (HC), whereas the other group had a rack, installed on the pen wall, containing horizontally placed wood logs (WL). WL are supposed to be more attractive and manipulable than hanging chains; these latters are frequently used as enrichment tools for pigs reared on a slatted floor. Animals were kept in collective pens, on a partially slatted floor and fed the same, rationed liquid diet. At slaughter (160 kg BW), all carcasses were assessed for pleurisy, pneumonia, pericarditis, white spots on the liver and gastric lesions. To this aim, stomachs were collected, opened and lesions in the pars oesophagea were observed, classified and scored (on a 0 to 6 scale) as: absent or mild (intact epithelium to mild hyperkeratosis: points 0-1); moderate (hyperkeratosis and mild erosions; points 2-4); or severe (moderate to severe erosion, and/or oesophageal ulcers or stenosis; points 5-6). Mann Withney U-test and Chi-squared test were used. The prevalence of pulmonary, hepatic, hearth and gastric lesions did not differ between the experimental groups. Although the absence of significance for overall severity of gastric lesions (3.0 vs 3.4 points in HC and WL group, respectively; $P>0.05$), highly significant differences were detected for the different classes of severity ($P<0.01$). Lesions were, in fact, distributed as follows: in HC group, 35% were classified as mild, 35% as moderate and 30% as severe; in WL group, 20% were classified as mild, 60% as moderate and 20% as severe. The reduced prevalence of severe lesions observed in WL group may indicate a higher welfare and health level that needs to be confirmed by further investigations presently in progress. Although gastric ulcers can represent an additional parameter to detect chronic welfare issues of pigs, it should be noted that, due to hygienic reasons (and contrary to pulmonary, hepatic and hearth lesions), the assessment of gastric lesions needs a dedicated, separate space and it cannot be presently regarded as a routine survey at slaughtering. This research was supported by Progetto AGER, grant n° 2011-0280

Development of measures of human-animal interaction for working equids

A.F. Brown

The Brooke, 41-45 Blackfriars Road, London SE1 8NZ, United Kingdom; ashleigh@thebrooke.org

The relationship between a domesticated animal and its primary human contact constitutes an integral component of the animal's daily experience, and has the potential to affect both physical and psychological welfare parameters. Direct human-animal interaction (HAI) is particularly prevalent in the working equid context due to frequent and prolonged periods of handling; however, few standardised measures of HAI for working equids have been defined. This study aimed to develop and pilot HAI indicators for equids involved in tourism in Petra, Jordan in order to assist the Brooke with identification of deficits and changes in owner behaviour towards their animals, and with evaluation of the need for, and impact of, project activities aimed at improving this. Measures of physical interaction, vocal interaction and oral/nose contact were recorded from riding donkeys/mules (DM), riding horses (RH) and carriage horses (CH) as they passed through predetermined observation zones along their usual working route. Observation zones were purposively selected on the basis of being likely locations of potential HAI; DM and CH each had two observation zones, and RH one. Data were recorded on two days in both low and high tourist seasons; in each, estimated average daily numbers of equids were 100 DM, 70 RH and 30 CH. Two of every three equids passing through the zone were recorded over two hours. As it was not feasible to identify individual subjects, some may have been recorded more than once during observation periods. The numbers of times HAI was observed in low and high seasons respectively were: 41 and 72 for DM; 25 and 44 for CH; 19 and 31 for RH. Observations of negative physical interaction (goading/beating) increased from low to high season from 12% to 35% in DM; 0 to 10% in CH; and 11% to 13% in RH. Observations of negative vocal interaction (shouts, harsh vocals) increased from low to high season from 2% to 10% in DM and 4% to 14% in CH. Observations of excessive oral pressure increased from low to high season from 16% to 39% in CH and 0 to 7% in RH. Whilst these data cannot be assumed to be reflective of the entire range of HAI, they suggest that these measures can provide a standardised means of assessment of HAI adequate to identify aspects of concern and detect trends. Such data can inform educational activities with owners and contribute to their impact assessment, thus demonstrating the utility of incorporating standardised HAI measures into welfare monitoring systems for working equids. Similar measures will now be developed for working equids in brick kilns, along with investigation into correlating HAI and animal-based welfare measures. This work could also be applied to other working animal contexts or animal systems with high levels of direct HAI.

Behavioural cues and personality as non-invasive indicators of alimentary stress in horses

A. Destrez[1], P. Grimm[1], F. Cézilly[2] and V. Julliand[1]
[1]AgroSup Dijon, 26, Bd Docteur Petitjean, 21079 Dijon, France, [2]UMR CNRS/uB Biogéosciences, 6, Bd Gabriel, 21079 Dijon, France; alexandra.destrez@agrosupdijon.fr

The digestive system of horses is adapted to high-fibre diet consumed in small amounts over a long time. However, during training, low-fibre diets are usually fed and may induce intestinal microbial disturbances and intestinal pain. Such an alimentary stress, due to microbial composition changes of caecal or colonic ecosystem, is usually assessed by invasive methods in experimental studies. The aim of the present study was to investigate to what extent changes in behaviour and personality are associated with alimentary stress. We discuss the possibility to use behavioural cues as a non-invasive technique to assess alimentary stress in horses. Six fistulated horses were used. The alimentary stress was a modification of diet from a high-fibre diet (100% hay) to a progressive low-fibre diet (from 90% hay and 10% barley to 53% hay and 47% barley in 5 days). During the diet of 53% hay and 47% barley, caecal and colonic total anaerobic, cellulolytic, amylolytic and lactate-utilizing bacteria were enumerated. Horses were observed during a 18-h time budget and two personality traits were measured: reaction to the presence of a novel object placed near a feeder in a test arena (neophobia) and reaction to an unfamiliar horse in a stall (sociability). The 18-h videos were analysed by scan sampling every 10 min using the following ethogram: lying, resting, feeding and being vigilant. Durations of feeding and being vigilant in novelty test and durations of being vigilant and interacting with the unfamiliar horse in sociability test were measured. Spearman correlations and adjusted p-values (Holm's method) were performed between data of microbial compositions and behaviours in tests. Our alimentary stress induced significant increases of colonic amylolytic (P=0.03) and lactate utilizing bacteria concentrations (P=0.006). According to the literature, these intestinal changes may lead to intestinal pain in horses. Duration of vigilance in sociability test was significantly correlated with caecal and colonic amylolytic bacteria concentrations (R^2>0.9, P<0.05). Duration of vigilance in novelty test tended to be correlated with caecal lactate-utilizing and with colonic amylolytic bacteria concentrations (R^2>0.9, P<0.08). Vigilance is considered as a negative emotional response and may reflect a negative state such as intestinal discomfort. These findings suggest that behavioural cues and personality traits may be used as non-invasive indicators of alimentary stress and, thus, might prove useful to prevent intestinal pain in farms. Further investigations are necessary to assess this link between welfare and microbial composition in the equine caecum and colon.

Behavioural measures during the sensitive period, as tool to assessed welfare in small ruminants

E. García y González[1], H. Hernández[1], E. Nandayapa[2], P. Mora[2], M.J. Flores[3], J.A. Flores[1] and A. Terrazas[2]
[1]Centro de Investigación en Reproducción Caprina-UAAAN., Torreón, Coahuila, México, 27054, Mexico, [2]Facultad de Estudios Superiores Cuautitlán, U.N.A.M., Cuautitlán Izcalli, 54714, Mexico, [3]I.N.I.F.A.P, Campo Experimental Zacatecas, Zacatecas, 98500, Mexico; angeterr_2000@yahoo.com.mx

In small ruminants the high mortality rate occurs during the first week after birth. The newborn mortality rate had been considered as a welfare measure in these farm animals. Sheep and goats gave birth to precocious newborns and immediately after birth those species have a sensitive period which they make maternal behaviour vulnerable, in terms that depend from physiological and sensorial factors that occur during this period. Many studies performed to evaluate the maternal temperament in both species, has developed a standard measures that could be useful to obtain a maternal score and predict the viability of the newborns. These measures have been made to evaluate some factors that affect the development of mother-young filial bond such as maternal nutritional, breed of the mother (in the case of sheep), maternal experience and others. In this sense, either in sheep or goats behavioural measures are: (1) presentation or not of dystocia; 2) latency to lick the newborn; (3) latency of the young to stand up; and (4) latency of the young to reach the udder. These parameters could be the first steps to evaluate the mother-young filial bonding in this species. Subsequent measures as a newborn´s temperature, maternal selectivity and mother-young distal recognition, are secondary parameters that could predict the viability of the young and the bonding consolidation. Additionally, the mentioned measures are easy to obtain, and their statistical analysis could be performed without problems. In this order of ideas, it´s been assessed that when a dystocia is presented, it will could a delay in the subsequent events, that could complicate the filial bonding process. Even more, if the mother get exhaust after a longer dystocia, it is possible that the sensitive period had been finish which decreases the possibilities that mother get imprint to the young. The dystocia could be analysed by a percentage or proportion of newborn that were assisted on the total young born. Other examples are maternal nutrition during pregnancy, thus many works done in sheep, and recently in goats, have showed that underfed mothers during pregnancy, display a poor maternal behaviour. Additionally, the behaviour of the young is also affected, in general a delay of the locomotors activities related with udder seeking. We concluded that those behavioural measures could be suitable to evaluate the welfare of the lambs and kids during the first week of life. GRANT: UNAM-DGAPA-PAPIIT-IN217012, FIS B/3872-1.

Poster 30

Nociceptive threshold testing in 6-month old *Bos indicus* calves following surgical castration

G.C. Musk[1], T.H. Hyndman[1], M. Laurence[1], S.J. Tuke[2] and T. Collins[1]
[1]Murdoch University, Murdoch, 6150, Australia, [2]University of Adelaide, Adelaide, Adelaide, Australia; g.musk@murdoch.edu.au

In Australia, analgesia is not routinely administered to cattle undergoing surgical castration on the farm. Furthermore, the pain associated with this husbandry practice has not been adequately characterised. To investigate analgesic strategies for this procedure, mechanical nociceptive threshold testing (MNTT) was used to assess the efficacy of lignocaine (L) and meloxicam (M) administered in the perioperative period. A decrease in the nociceptive threshold (NT) from baseline indicates the development of hyperalgesia and an increase implies analgesic efficacy. It was hypothesised that the NT would increase following administration of 0.5 mg/kg subcutaneous (SC) M and 2 mg/kg of SC and intra-testicular L given pre operatively. 48 Brahman bull calves were randomly divided into six equal groups: no surgery control (NC); surgical castration (C) without analgesia; C and M_{pre-op}; C and $M_{post-op}$; C, L and $M_{post-op}$; C and L. MNTT was performed the day before surgery (day -1) and on days 1, 2, 6, 10 and 13 after surgery. A handheld manual pneumatic device (ProdPro, Topcat Metrology Ltd.) with a 1 mm blunt pin was used to deliver a mechanical stimulus to a maximum of 27 N lateral to the sacrum. The operator stood on a raised platform next to a race which held six randomly selected animals. The NT was recorded when a response to the stimulus was observed. Responses included stepping away from the stimulus, kicking, tail swishing or lifting the leg closest to the site of the stimulus. Each test was performed five times with at least five minutes between each test. The mean of the five tests were used for analyses. Data were analysed with a mixed effect linear model with NT as the response variable and day and analgesic treatment as predictors (P<0.05 was considered significant). Data are expressed as mean (SD). For all groups, there was a trend toward decreasing NT over the study period but there were no significant differences between groups. The NT on days -1 and 1 were 22.1 (4.2) N and 19.9 (4.5) N for NC; 20.7 (4.5) N and 20.4 (5.4) N for C; 21.8 (6.4) N and 20.1 (5.5) N for C and M_{pre-op}; 22.6 (4.9) N and 17.5 (5.1) N for C and $M_{post-op}$; 21.8 (4.3) N and 21.4 (5.4) N for C, L and $M_{post-op}$; and 20.3 (5.0) N and 16.9 (6.2) N for C and L. MNTT for assessment of analgesic efficacy in *Bos indicus* bull calves requires further refinement. The temperament of the animals, the proximity of the MNTT device operator to the animal, the site of stimulus application and the type of stimulus may impact on the collection of meaningful data. Given the usefulness of MNTT in other species, further work is warranted.

Effects of housing area on pecking behaviour and plumage quality of free-range layer chickens

M. Petek, E. Topal and E. Cavusoglu
Uludag University, Zootechnics, Department of Zootechnics, Faculty of Veterinary Medicine, University of Uludag, 16059 Bursa, Turkey; petek@uludag.edu.tr

This study was made to investigate the effects of housing area on pecking behaviour and plumage quality of free range layer chickens. Six hundred 16-weeks of age layer pullet (Lohman Brown) obtained from a commercial company were reared in an experimental free range house. The experimental house were divided in three similar parts as slatted, litter and range area. The birds were allowed to range area first time at 22 weeks of age. Plumage quality and pecking behaviour of the birds were measured for eight weeks intervals from 24 to 60 weeks of age. Total feather damage increased with age in all housing area, significantly ($P<0.001$). Compared to slatted and litter area, the birds in range area had less feather damage. The feather score of tail body region in all groups was greater (more damage) than the others, significantly ($P<0.001$). Incidents of gentle feather pecking in every period was more frequent while aggressive pecking was less. As conclusion it can be said that housing area had a significant effect on pecking behaviour and feather damage of the birds in free-range housing system.

Welfare criteria for on-farm slaughter of cattle when using a rifle as a stunning/killing method

M. Von Wenzlawowicz[1], S. Retz[2] and O. Hensel[2]
[1]bsi Schwarzenbek, Grabauer Straße 27A, 21493 Schwarzenbek, Germany, [2]University of Kassel, Faculty of Organic Agricultural Sciences, Department of Agricultural Engineeri, Department of Agricultural Engineering, Nordbahnhofstraße 1, 37213 Witzenhausen, Germany; mvw@bsi-schwarzenbek.de

On-farm slaughter using a rifle has become more popular in Germany since hygienic rules for slaughtering cattle, raised outdoor, had been changed in October 2011. Due to the fact that only few legal requirements exist for this stunning/killing method and anecdotal experience on failures is high, investigations have been carried out to gather information about best practice and how to avoid welfare infringements. To define suitability of guns and ammunition for stunning and killing of adult cattle, isolated heads of electrically stunned cattle (German Angus, Galloway and crossbreeds, total n=44) were shot and sections of the skulls were prepared and subsequently examined with regard to shot location, shooting angle, penetration depth and the traumatised brain tissue. Optimum shooting position and angle as well as ammunition were defined. In a second study, cattle (German Angus, Galloway and Highland Cattle, total n=31) were shot according to the recommendations found, stunning effectiveness was assessed and a rough pathological investigation of brain damage was performed. Effectiveness of stunning and killing was monitored using the following symptoms: immediate collapse, no regular breathing, no eye movements. Ammunition for rifles was proven to be suitable when calibre .22 (Magnum, semi-hollow) was used on a distance up to 30 m. Bigger calibre and stronger cartridges had no benefits for animal welfare. The optimum shooting position was frontal on the head 2 cm above the crossover point of two imaginary lines drawn between the base of the horns and the contralateral eye. Lack of precision lead to the bullet missing the brain cavity and consequently impaired stunning effectiveness. Shooters capabilities are crucial to avoid failures. Shooters have to be well trained and should prove their aiming accuracy before each shot. Other important criteria are the preparation of the flock and a special shooting area with elevated position for the marksman and backstop. Shooting area has to be restricted to have fast access to the animals after the shot and avoid flight possibilities. Animals should be habituated to the shooting area and not be shot alone. Only if all the above mentioned criteria are met, on farm slaughter by rifle should be permitted.

Assessing animal welfare in dairy calves: at herd level and across age groups

M. Reiten[1], T. Rousing[1], N.D. Otten[2], H. Houe[2], B. Forkman[2] and J.T. Sørensen[1]
[1]Aarhus University, Department of Animal Science, Blichers Alle 20, 8830, Denmark, [2]University of Copenhagen., Department of Large Animal Sciences, Gronnegaardsvej 8, 1870, Denmark; marireiten1@gmail.com

On-farm welfare assessments on dairy cattle herds often focus only on the dairy cows. However, calves 0-180 days represent 10 to 20% of all the animals in a typical dairy cattle herd and there are several animal welfare problems that may occur among dairy calves. Therefore there is a need for developing a welfare assessment system for quantifying animal welfare for calves 0-180 days of age. The aim of this study is to develop a protocol of relevant welfare variables, a sampling strategy and an index concept to quantify and assess animal welfare for dairy calves from birth to 6 month of age at herd level. The protocol is based on a definition of animal welfare as an affective state. The welfare assessment protocol include: (1) a clinical assessment of the calf (i.e. body condition score, lesions, cleanliness wounds from inter-sucking, umbilical inflammations, diarrhoea, hampered respiration, nasal discharge, hairless patches, hear coat appearance); (2) information on housing and management (i.e. milk feeding, roughage feeding, concentrate feeding, water provision, space allowance, bedding material and quality, humidity, cleanliness of pen, group composition, physical contact and access to artificial teat); and (3) register data on mortality and disease treatments. The protocol is used to assess animal welfare in 36 Danish dairy herds with more than 200 dairy cows. In each herd 40-60 calves are assessed. Half of the herds are conventional dairy herds and half of the herds are organic dairy herds. The clinical assessment and information on housing and management is conducted on a one day visit and the mortality and treatment data cover a 3 month period prior to the visit. The animal welfare at farm level is transformed into an index for each of the age groups: 1-14 days, 15-90 days and 91-180 days. The herds are ranked according to each index and the variation in ranking between the 3 age groups is analysed according to the different indexes. The difference in the welfare assessment index sensitivity between the 3 age groups for identification of differences in animal welfare between the two production systems (conventional versus organic) will also be analysed. The results will be presented at the conference.

Interobserver reliability of the Welfare Quality® animal welfare assessment protocol

I. Czycholl[1], C. Kniese[2], L. Schrader[2] and J. Krieter[1]
[1]*Institute of Animal Breeding and Husbandry, Christian Albrechts University, Olshausenstr. 40, 24118 Kiel, Germany,* [2]*Institute of Animal Welfare and Animal Husbandry, Friedrich Loeffler Institute, Doernbergstr. 25/27, 29223 Celle, Germany; iczycholl@tierzucht.uni-kiel.de*

To test the feasibility and reliability of the Welfare Quality® Animal Welfare Assessment Protocol for Growing Pigs (AWAP) under practical conditions, an Interobserver Reliability study was performed by 3 trained observers, whereby 19 combined assessments were fulfilled by observers 1 and 2 and 10 by observers 1 and 3. A Qualitative Behaviour Assessment (QBA), direct behaviour observations (BO) using instantaneous scan sampling, a Human Animal Relationship Test (HAR) and checks for different individual parameters, e.g. presence of tail biting, wounds and bursitis were applied. The observations took place at the same time and on the same animals, but completely independent from each other. The following statistical agreement parameters were calculated: Spearman Rank correlation Coefficient (RS), Intraclass Correlation Coefficient (ICC) and Smallest Detectable Change (SDC). In the QBA, no agreement was found in any of the terms (e.g. active: RS 0.33, ICC 0.17, SDC 0.3; fearful: RS 0.14, ICC 0, SDC 0.25) when considering the direct values of the scale in mm. For further analysis, a Principle Component Analysis was carried out. The BO showed good to moderate agreement in the ordering of pigs into the behavioural categories (social: RS 0.45, ICC 0.5, SDC 0.05, negative social: RS: 0.46, ICC 0.41, SDC 0.01, use of enrichment material: RS: 0.75, ICC 0.75, SDC 0.04, pen investigation: RS: 0.61, ICC 0.61, SDC 0.08). The HAR was carried out twice with the same animals as the observers entered the pens one after the other to minimise mutual interference and to not affect the pigs by 2 intruders. However, this diminished the informative value about the Interobserver Reliability as the reaction of the pigs towards the second person was influenced by the first which may explain the rather low agreement (RS 0.47, ICC 0.5, SDC 0.14). For the individual parameters, varying agreement was found: In the grading of tail biting (RS 0.52, ICC 0.47; SDC 0.02) and wounds (RS 0.43, ICC 0.56, SDC 0.03) the observers agreed well. However, the parameter bursitis showed great differences (RS 0.45; ICC 0.27, SDC 0.14). This can be explained by the practical experience that bursitis is difficult to assess when the animals are moving fast or the legs are dirty. Some parameters did not occur at all or very rarely (e.g. rectalprolapse), which makes an assumption about their reliability impossible. In the ongoing research, a Retest study will be analysed and video observations will be linked to the BO of the AWAP.

Study on hair cortisol variations during two successive reproductive cycles in sows

E. Nannoni, N. Govoni, F. Scorrano, A. Zannoni, M. Forni, G. Martelli, L. Sardi and M.L. Bacci
University of Bologna, Department of Veterinary Medical Sciences, Via Tolara di Sopra 50, 40064
Ozzano Emilia (BO), Italy; eleonora.nannoni2@unibo.it

The assessment of cortisol concentration in hair has been proposed as a minimally-invasive technique capable of furnishing information over medium/long period stress response. The aim of the present work was to assess the reliability of using bristles as a matrix for the assessment of mean cortisol level in swine. To this aim, bristles were collected by shaving the rump region of 30 naturally synchronized sows housed in a commercial breeding farm, over two successive reproductive cycles. Samples were collected at three key physiological phases (before delivery-BD; weaning time-WT; pregnancy diagnosis-PD) during two consecutive reproductive cycles (1 and 2), which naturally developed in different seasons of the year (PD1 occurred in April, BD1 in June, WT1 in July, PD2 in September and BD2 in November). Samples were processed as follows: hair was repeatedly washed in water and isopropanol to remove the external debris. The sample was then dried, pulverized and 60 mg were weighed. Methanol was added to the sample, and the vial was incubated overnight for steroid extraction. Methanol was then transferred to another vial, evaporated and stored at -20 °C pending cortisol assessment, which was carried out by a validated RIA. Differences in cortisol levels between the sampling points were studied with one-way ANOVA and the post-hoc Duncan multiple range test. Two-way ANOVA was applied to evaluate the effects of reproductive phase and season. PD1 and PD2 values were 20.10±1.95 and 16.29±2.15 pg/mg, respectively and were significantly higher (P<0.001) when compared to values obtained at BD and WT. Mean cortisol value at BD2 was higher (P<0.001) than at BD1 (10.48±0.96 vs 5.17±0.51 pg/mg). Our results showed a significant effect of the reproductive phase on cortisol level in bristles (P<0.00001), suggesting, in agreement with what has been well described in the literature, a higher activation of the adrenal response during late pregnancy and lactation with respect to early-mid pregnancy. Such a difference could be not only due to physiological hormonal changes, but also to the different housing conditions (single crates vs group housing). An effect of season was also detected, with the lowest cortisol values being recorded during the hot season (P<0.005). However, further research would be necessary to investigate if this reduction in cortisol levels during the hot season plays a role in swine seasonal infertility. Our data suggest that cortisol from hair could be regarded as a reliable method for gathering information, otherwise very difficult to collect, on the mean cortisol levels of pigs over a medium/long period of time.

A preliminary assessment of horse's welfare on breeding and fattening farms in Japan

S. Ninomiya
Gifu University, Faculty of Applied Biological Sciences, 1-1 Yanagido Gifu, 501-1193, Japan;
nino38@gifu-u.ac.jp

Welfare quality becomes an important aspect of animal production and assessment protocols of animal welfare have also been developed. We conducted a preliminary welfare assessment of horse mainly for breeding and fattening in Japan. 86 fattening horses (F), 36 breeding mares (B), 16 stallions (S) and 59 weanling horses (o) were used. Their breeds were Breton, Belgian, Percheron or mixed breeds with them. They were housed in 6 horse farms. As a welfare assessment, their body condition score (BCS), body's cleanliness, hoof's condition, behavioural responses to frustration after feeding and avoidance response to human approach (AvoR) were observed. BCS of each animal were recorded referring to Henneke *et al.* The cleanliness of each animal's body was categorized as score 3 (no dirt), score 2 (no more than 10% of body's surface which contacts with ground during sternum lying is dirty) or score 1 (more than 10%). Hoofs of each animal was categorized as score 3 (normal), score 2 (hoofs were growing a little) or score 1 (hoofs were growing). Behaviours related to frustration (self-grooming, pawing, bedding investigation, coprophagia, licking or biting a stall structure) were observed using a scan sampling technique at 1 min intervals for 2 hours after feeding. Measuring AvoR, an assessor stood at a distance of 3 m in front of each animal during feeding and approached it at a speed of one step per second. AvoR of each horse was categorized as score 3 (being able to touch), score 2 (a horse avoid being touched by an assessor but it do not run away) or score 1 (a horse avoid being touched by an assessor and run away). Each score or a total time budget of behaviours related to frustration were averaged for rearing categories (F, B, S, O) of each farm. The scores were compared among rearing categories by Kruskal-Wallis test. There were some differences among rearing categories in BCS, behavioural responses to frustration after feeding and AvoR ($0.05 < P < 0.10$). These results indicated that rearing methods might affect the horse's welfare and these scores would be useful for welfare assessment of horses on breeding and fattening farms in Japan.

Indicators of thermal comfort for on-farm welfare assessment in dairy goats

M. Battini, L. Fioni, S. Barbieri and S. Mattiello
Università degli Studi di Milano, DIVET, via Celoria 10, 20133 Milan, Italy; monica.battini@unimi.it

Up to date, scarce information is available on valid and feasible indicators of thermal comfort in goats. Aim of this investigation was to test the validity and feasibility of indicators for on-farm assessment of cold and heat stress in dairy goats. The study was performed in three pens (n. of animals/pen: 233, 97, 71) of intensively farmed dairy goats in Italy. For each pen, data were collected in the morning (8:00), afternoon (12:30) and evening (17:00) in January, April-May and July 2013, for a total of 9 observation sessions/pen (3 time bands per 3 days). During each session, temperature (°C), relative humidity (%) and wind speed (km/h) were recorded and Thermal Heat Index (THI) was calculated. The sessions were allocated to three climatic seasons, depending on THI ranges (min-max): Cold (32-45), Neutral (54-64) and Hot (66-83). Both cold and heat stress were scored by visual assessment from outside the pen. For cold stress, we used a 3-point scale scoring system (shivering score): no sign of stress (score 0), moderate stress with bristled hair on the back (score 1), high stress with shiver (score 2). For heat stress, we used a 3-point scale scoring system (panting score): normal respiration (score 0), elevated respiration with closed mouth (score 1), panting with open mouth (score 2). Shivering score 2 was recorded only in one goat in Cold season and panting score 2 only in two goats in Hot season, therefore scores 1 and 2 were aggregated for statistical analysis. The frequencies of goats suffering from either cold or heat stress were compared between seasons using a Chi-Squared or Fisher's Exact Test, where appropriate. The percentage of goats showing signs of cold stress was significantly higher in Cold season than in Neutral (P<0.01) and Hot (P<0.001) season (1.77, 0.38 and 0.00%, respectively). In Neutral season, cold stress was observed only during a morning session with low temperature (12 °C) and high humidity (93%), whereas in Cold season it was equally distributed during the three time bands. Signs of heat stress were recorded only in Hot season in 11.73% of goats (P<0.001 compared to both Neutral and Cold season). Our results show that the shivering and the panting scores used in this study are valid indicators to detect thermal stress in goats. The visual assessment from outside the pen confirms the on-farm feasibility of both indicators. A further refinement of the scoring systems may consider merging scores 1 and 2, thus obtaining a 2-point scale scoring system, useful also for farmers to gather early information about thermal comfort of goats. The authors thank the EU VII Framework program (FP7-KBBE-2010-4) for financing the Animal Welfare Indicators (AWIN) project.

Multiparametric approach to assess pain in sheep following an experimental surgery

M. Faure[1], V. Paulmier[1], A. Boissy[1], A.S. Bage[1], C. Ravel[1], A. Guittard[2], A. Boyer Des Roches[3] and D. Durand[1]
[1]INRA, UMR 1213 Herbivores, Saint-Genès-Champanelle, 63122, France, [2]INRA, UE1354 Ruminants de Theix, Saint-Genès-Champanelle, 63122, France, [3]Université de Lyon, VetAgro Sup, UMR 1213 Herbivores, Marcy l'Etoile, 69280, France; denys.durand@clermont.inra.fr

Surgeries are usually performed in farm animals both for routine practices (e.g. caesarean, castration…) and experimental aims (e.g. implementation of digestive cannulas, of blood catheters…), and potentially elicit sources of pain. However pain is a complex process and its assessment in farm animals is not fully well documented. Providing practical tools to assess pain in livestock are essential not only to meet societal expectations but also to ensure farmers, veterinarians and scientists in their attempts to reduce animals' pain. Few works has been realised for assessing pain in routine practices but little has been done in experimental surgery. The objective of the present study was to determine in sheep the relevance of a multi-parametric approach to assess pain following an experimental surgery putatively invasive. Two groups of six Texel castrated male sheep were used. Group S underwent an implantation of three cannulas (in rumen, proximal duodenum and terminal ileum) and the associated constraints (i.e. fasting, anaesthesia and peri-operative drugs treatment), while group C ('Control') underwent neither surgical intervention nor the associated constraints. Behavioural patterns (i.e. general activity; antalgic postures…) that allowed calculation of a synthetic score were recorded 1 week before and during 2 weeks after the surgical intervention. Blood samples were taken to determine plasma cortisol and haptoglobin indicators of stress and inflammatory status. Statistical analysis was performed by a variance-covariance analysis (GLM procedure on SAS) in which the variance of the period following experimental treatments was calculated taking into account the period before treatment as a covariate. Before the surgery, basal values (± SD) were not different between the two groups (plasma cortisol: 7.4±6.2 ng/ml; plasma haptoglobin: 0.2±0.06 mg/ml; behavioural score: 4±5). After the surgery, indicators values were higher in animals of group 'S' than that of group 'C' ($P<0.05$) with an increase of +150, +297 and +320% for these respective indicators. In conclusion, the multi-parametric approach discriminates sheep underwent surgery those not underwent surgery. However in this study, it was difficult to discriminate whether these indicators were related to surgery act per se or to surgery and the associated constraints (i.e. fasting, anaesthesia and peri-operative drug treatments).

Salivary s-IgA as an indicator of positive emotions in cattle?

S. Lürzel[1], J. Götz[1], A. Stanitznig[1], M. Patzl[2] and S. Waiblinger[1]

[1]Institute of Animal Husbandry and Animal Welfare, Vetmeduni Vienna, Veterinärplatz 1, 1210 Vienna, Austria, [2]Institute of Immunology, Vetmeduni Vienna, Veterinärplatz 1, 1210 Vienna, Austria; stephanie.luerzel@vetmeduni.ac.at

Positive experiences are essential for a high 'quality of life'. It is therefore crucial to find indicators for positive emotions. Secretory immunoglobulin-A (s-IgA) in saliva has been proposed as a physiological marker, as it has been shown to increase in humans when positive emotions are induced. The aim of our study was to investigate salivary s-IgA concentrations and cortisol in dairy cows on the day they are moved to pasture after a long period of loose housing. We hypothesized an increase in s-IgA on the pasture. Ten experimental cows of the lactating cow herd of the Vetmeduni Vienna (cubicle loose housing with an outdoor run) were habituated to saliva sampling. Samples were taken on the first day of pasturing (Day 0, 9 samples), one day before (Day -1, 4 samples), and one day a week later (Day 8, 4 samples). Behaviour was recorded via direct observation using instantaneous sampling, four times 15 min on Day 0 as well as on Day 8. The cows showed more play behaviour directly after being moved to the pasture than during all later observations (Wilcoxon test: n=9, P<0.05, Bonferroni corrected) and on Day 0 than Day 8 (n=8, P=0.012). There was a trend towards an increase in s-IgA levels when comparing the samples taken directly before with the ones taken after moving the cows to pasture (n=9, P=0.08, one-sided) due to an increase in six out of nine cows, while cortisol increased significantly (n=9, P=0.004). Mean s-IgA concentration on Day -1 was higher than on Day 0 (n=9, P<0.05) and on Day 8 (n=8, P<0.05), and variability within cows was quite high. Cortisol decreased from Day -1 to Day 8 (n=8, P<0.05). Cortisol and IgA showed a weak positive correlation. Moving the cows to pasture did elicit play behaviour as expected and thus, the cows most likely experienced positive emotions. The trend towards an increase in s-IgA after the cows entered the pasture barely supports its usefulness as a marker for positive emotions; the rise in cortisol might be due to arousal and/or high locomotor activity. However, lower mean IgA on pasture paralleled by lower cortisol argues against a general association with positive affective states. Furthermore, recent studies in humans have shown that s-IgA may also increase following a negative stimulus. In sum, salivary IgA does not seem to be a promising indicator for positive emotions in cattle, but further studies are necessary to investigate its sensitivity and specificity in a larger sample and in diverse situations.

Is the duration of lying down movements associated with the daily lying time in dairy cows?

L. Mounier, A. Dufaut, E. Delval and I. Veissier
Inra, 1 avenue Bourgelat, 69280 Marcy l'étoile, France; luc.mounier@vetagro-sup.fr

Lameness is probably the most welfare problem facing the dairy cows. A low lying time is one of the most important environmental risk factor for lameness. Assess the lying time in a herd is very important to reduce the prevalence of lameness. Several indicators have been developed to assess the lying behaviour: total lying time, frequency and duration lying bouts. Recording such indicators are time consuming and then not suitable for on-farm assessments. Automatized measurements of lying behaviour using data loggers have been developed but stay technically difficult to perform on farm. Other indicators, like the cow comfort index or the stall use index are easier to perform but were showed to be not associated with the daily lying time. In the welfare Quality® protocol, comfort around resting is assessed using the time needed to lie down. Previous work has shown that cows tend to spend more time lying on softer surfaces and that cows lie down less often when less bedding is used, perhaps because of the considerable weight placed on the knees during the transition from standing to lying. We could suppose that the time needed to lie down could be used as an indicator of the total daily time and the frequency and duration of lying bouts. Two experiments were performed to assess the association between the duration of lying down movements and the lying behaviour. The first one was conducted at herd level on 22 dairy commercial farms and the objective was to assess the association between the average lying behaviour and the average time needed to lie down. The second one was conducted at individual level on one experimental farm and the objective was to assess if the time needed to lie down for a cow was associated to its lying behaviour. Lying behaviour was recorded using electronic data loggers that were attached on the cows. The duration of lying down movements was recorded by direct observations on commercial farms and by video on the experimental farm. At herd level, the average duration of the lying down movement was 5.56 s and the cows spent 642.36 min lying with an average of 8.72 bouts. At individual level, the average duration of the lying down movement was 5.45 s and the total lying time was 695.06 min. Whether at herd or herd level, the duration of lying down movements was not correlated to the total time spent lying, neither to the frequency and duration of lying bouts. Moreover, the duration of lying down movements was very variable within farms, making this indicator difficult to use on farm. Our results show that the total lying time could not be assessed by the duration of lying down movements and dairy cows and that other indicators should be developed.

Poster 41

Is it appropriate to measure plasma cortisol for assessing prepartum sow welfare conditions?

J. Yun, K. Swan, O.A.T. Peltoniemi, C. Oliviero and A. Valros

University of Helsinki, Department of Production Animal Medicine, P.O. Box 57, 00014 University of Helsinki, Finland; pilot9939@gmail.com

Sow plasma cortisol concentrations have been widely used for measuring their welfare conditions. We investigated whether provision of nest-building materials and space prior to parturition could reduce prepartum sow plasma cortisol concentrations as an indicator of prepartum stress, whether cortisol concentration could affect oxytocin concentrations in prepartum sows, and whether this assessment could be appropriate to measure prepartum sow welfare conditions. We allocated 33 sows in: (1) CRATE: the farrowing crate closed (210×80 cm), with provision of a bucketful of sawdust; (2) PEN: the farrowing crate opened, with provision of a bucketful of sawdust; (3) NEST: the farrowing crate opened, with provision of abundant nest-building materials. We collected sow blood samples for cortisol and oxytocin assays via indwelling ear vein catheters on days -3, -2, -1 from parturition twice a day. Cortisol and oxytocin concentrations were analysed with a mixed model using repeated measures. Pearson correlation coefficients (r) were used to examine interactions of oxytocin concentrations between prepartum and parturition. Cortisol concentrations were tested to determine interactions with oxytocin concentrations by regression analysis. Sow plasma oxytocin concentrations in NEST were significantly greater than in CRATE during the prepartum period (P<0.05). Prepartum oxytocin concentrations were significantly correlated with farrowing oxytocin concentrations (r=0.64, P<0.0001). However, prepartum sow plasma cortisol concentrations were not significantly different among three different farrowing environments (P>0.10), and had no interactions with sow prepartum or parturition oxytocin concentrations. These results indicate that provision of nest-building opportunity prior to parturition could increase prepartum sow plasma oxytocin concentration, possibly through allowance of natural behaviour, but did not affect to cortisol concentrations in prepartum sows.

Is it possible to measure the welfare performance of animals in diverse farming systems?

C.E. O'Connor[1], L.J. Keeling[2] and M.W. Fisher[1]

[1]Ministry for Primary Industries, Animal Welfare Standards, P.O. Box 2526, Wellingotn 6140, New Zealand, [2]Swedish University of Agricultural Sciences, Department of Animal Environment and Health, Box 7068, 750 07 Uppsala, Sweden; cheryl.oconnor@mpi.govt.nz

Viable livestock farming requires practices that are not only productive, profitable and sustainable but also that fit with society's expectations on animal welfare. Transparent demonstration of how these expectations have been met will be paramount in the future for countries both exporting and importing animal products. Strategically one of the key routes to improved animal welfare is measuring animal welfare performance over time and communicating this information. Reporting animal welfare performance enables claims about good animal welfare to be backed up at the farm or country level and is also an important step to understanding where improvements might be required. We recognise that governments and sectors will need to work together to agree on the purpose of the assessment and the approaches to be used. Challenges include agreeing on performance indicators, deciding equivalence of methods and determining how to best collect and present measurement information. There is no one comprehensive, fully-validated system for on-farm welfare assessment that accommodates the diversity of species, production environments, animal management practices and people's expectations. If standard methods can be established then different farms, industries or countries can use the assessment according to their own needs and determine what is acceptable or not for their own situation, depending on the purpose of the assessment. For intensive systems, assessments based on combinations of health and production data together with observation of behaviour and physical appearance of animals within a group offer reliable and feasible indicators for the assessment of welfare. However can, or should, such measurements be extended to outdoor systems, e.g. extensive sheep and beef cattle? In this paper we discuss the advantages and disadvantages of converting indicators into performance measures to enable benchmarking of both extensive and intensive farming systems.

Poster 43

The assessment of animal welfare in pig research settings

M. Rice, C. Ng and L. Hemsworth
Animal Welfare Science Centre, The University of Melbourne, Parkville, 3010, Australia;
mrice@unimelb.edu.au

Increasing community concern about farm animal welfare has highlighted a need for a valid method of on-site welfare assessment. For a welfare monitoring system to be effective and acceptable by all key stakeholders, it must be able to measure broadly-accepted welfare indices in a way that is practical, reliable and repeatable. A number of livestock welfare monitoring schemes have been developed, both in Australia and abroad, across a range of livestock industries and while these welfare assessment tools have undeniable merit on farm, they require some adaptation if they are to be useful in accurately assessing welfare in a research setting. This Pork CRC funded project aimed to develop a protocol to assess the welfare of sows and their piglets in research. This will also enable the identification of those experimental treatments which while having positive effects on productivity, may have adverse effects on sow and piglet welfare and thus would not be acceptable to industry. A literature review of current animal welfare assessment tools identified 9 animal-based and 14 resource-based welfare criteria according to existing validations as well as the reliability and feasibility of the methods. The identified indices were tested for feasibility in research settings and adjusted as required. Using these adjusted indices, two welfare assessment questionnaires were developed and circulated for review to two research units running animal-based research projects. The questionnaires were refined based on this feedback. The final animal-based indices included fear of humans, body condition score, bursitis, lameness, skin injury scores, stereotypies, panting, manure on body and play behaviours. The resource-based indices used were ventilation, housing type, water supply, age of weaning, castration, ammonia level, mortality, morbidity, cleaning routine, tail docking, teeth clipping/grinding, bedding, environmental temperature, floor type, environmental enrichment, and space allowance. The identified welfare indices are refined and standardized, enabling the development of a reliable, practical and feasible protocol, which will be used to assess the welfare of sows and their piglets in husbandry and housing systems under study in Pork CRC funded research projects. Further research is required to identify the index scores within the acceptable ranges of welfare outcomes. The final outcomes of this project have the potential to be used both in future animal welfare research and in the development of animal welfare assessment protocols suitable for the use on-farm in Australia.

Anticipatory behaviour as an indicator of reward sensitivity in lambs: effects of prenatal stress

M. Coulon[1,2], A. Boissy[1,2], C. Mallet[1,2] and R. Moe Oppermann[3]
[1]*Clermont Université, VetAgro Sup, UMR Herbivores, BP 10448, 63000 Clermont-Ferrand, France,* [2]*INRA, UMR1213 Herbivores, Site de Theix, 63122 Saint-Genès Champanelle, France,* [3]*Norwegian University of Life Sciences, Production Animal Clinical Sciences, Box 8146 Dep, 0033 Oslo, Norway; marjorie.coulon@clermont.inra.fr*

The intensity of species-specific anticipatory behaviour in response to a conditioned stimulus (CS) signalling an attractive reward (unconditioned stimulus; US) is thought to reflect reward sensitivity. Such reward sensitivity could be an indicator of welfare state since it is affected by previous experiences of animals: following negative experiences the animal tends to increase its anticipatory behaviour. The aim of the present study was to validate the anticipatory behaviour as an indicator of both the reward sensitivity and welfare state in lambs. After having characterized anticipatory behaviour induced by trace conditioning, the effect of a prenatal stress on the intensity of anticipatory behaviour was studied. Fifty-four lambs were divided into four groups crossing two factors: lambs born from stressed ewes during pregnancy (daily repeated exposed to unpredictable and negative challenges such as social isolation, mixing, transport, delay feeding times) or from control ewes, and lambs were trained (Experimental lambs: EL) or not (Control lambs: CL) to anticipate reward delivery. EL were initially trained to associate a bell signal (CS) and concentrate (US), and then to anticipate by gradually increasing the CS-US interval up to 1 min after 5 days of training. CL were also exposed to the bell signal and concentrate but in a random order. All lambs were then tested using a CS-US interval of 1 min and observed continuously during 3 periods: 1min before CS, 1 min in the CS-US interval, and 1 min after US. The results demonstrated that the EL responded to the CS with an increase in the expression of anticipatory behaviour compared to CL. They increased their frequency of behavioural transitions (1.2 ± 0.5 to 5.1 ± 0.5, $F(2,114)=4.4$, $P=0.01$) and percentage of time spent in vigilance (1.3 ± 5.6 to $67.2\pm5.6\%$, $F(2,114)=8.8$, $P<0.001$) and decreased their percentage of time spent lying down (64.2 ± 5.8 to $7.2\pm7\%$, $F(2,114)=5.73$, $P<0.01$) and on maintenance behaviour, especially in EL born from stressed mother (18.6 ± 5 to $0.5\pm5\%$, $F(2,114)=15.93$, $P<0.001$). In addition, lambs born from stressed ewes spent more time in vigilance after the bell signal than lambs born from control ewes (60.1 ± 4.5 vs $38.6\pm5.7\%$, $P=0.04$). Anticipatory behaviour can thus be induced in lambs by using trace conditioning, and prenatally negative experiences can induce an increased sensitivity for reward. This study is promising for promoting anticipatory behaviour as a non-invasive welfare indicator in sheep.

Assessing animal welfare during long road journeys: a tool for certification

C. Pedernera[1], A. Velarde[1], W. Ouweltjes[2], K. Visser[2], S. Messori[3], B. Mounaix[4], P. Chevillon[5], M. Marahrens[6], E. Sossidou[7], P. Ferrari[8] and H. Spoolder[2]
[1]Institut de Recerca i Tecnologia Agroalimentàries, Animal Welfare Subprogram, Veinat de Sies s/n, 17121 Monells, Spain, [2]Wageningen UR Livestock Research, P.O. Box 65, 8200 AB Lelystad, the Netherlands, [3]Istituto Zooprofilattico Sperimentale dell'Abruzzo e del Molise, Campo Boario, 64100 Teramo, Italy, [4]Institut de l'Elevage, Service Santé et Bien-être des Ruminants, 35652 Le Rheu Cedex, France, [5]IFIP Institut du Porc, 3-5 rue Lespagnol, 75020 Paris, France, [6]FLI, Institute for Animal Welfare and Animal Husbandry, Doernbergstr 25-27, 29223 Celle, Germany, [7]Hellenic Agricultural Organization-DEMETER, Veterinary Research Institute, 57001, Thermi-Thessalon, Greece, [8]Research Centre on Animal Production, CRPA, 42121, Reggio Emilia, Italy; cecilia.pedernera@irta.cat

Animal transportation on long journeys, exceeding 8 hours, is a common practice in Europe. Current regulation for protection of animals during transport (Regulation EC 1/2005) is based on requirements related to resource based measures and management recommendations. The present study aimed to develop protocols for the welfare assessment of sheep, cattle, horse and pig transports to provide a foundation for a quality certification system. The work was part of the project 'Development of EU wide animal transport certification system and renovation of control posts in the European Union', funded by DG SANCO. Based on the four principles and 12 criteria of Welfare Quality®, four protocols were developed using animal based measures together with handling, resource, truck and transport measures. Assessments were carried out upon arrival, during unloading and in the resting pens. Pilot assessments were used to refine the protocols by selecting a set of suitable indicators applicable under field conditions. Protocol feasibility was then tested for 50 commercial transports per species, at control posts, slaughterhouses or assembly centres in five countries (France, Germany, Greece, Italy, and Spain). These assessments were performed by trained observers. Training included both theory and practice; assessors were trained until consensus of agreement was reached. It was concluded that the new protocols provided good inter-observer reliability and good feasibility, and are applicable for assessment of relevant animal welfare indicators during transport. It is recommended that future research should consider inter species differences, for example between calves and heifers.

Poster 46

Identifying individual differences in positive behavioural expressions in pigs

S.M. Brown[1], M. Klaffenböck[2] and A.B. Lawrence[1,3]
[1]Roslin Institute, University of Edinburgh, Easter Bush, Midlothian, EH25 9RG, United Kingdom,
[2]University of Natural Resources and Life Sciences, Agricultural Sciences, Gregor Mendel Straße 33, Wien, 1180, Austria, [3]SRUC, Roslin Institute, Easter Bush, Midlothian, EH25 9RG, United Kingdom; sarah.brown@ed.ac.uk

The debate over animal welfare has been largely focused on prevention of negatives. More recently the debate has moved more to discussing how to give animals positive experiences. Play behaviour is one such positive state which has the potential to be used as a proxy measure for positive welfare. Previous studies have suggested that play can improve the health status and survival of individuals and reduce recovery time after stressful events later in life. While the apparent long term benefits of play have received a lot of attention, the differences between individuals in the quality and quantity of play has been much less studied. Understanding individual or group differences in play behaviour is an important approach in developing play as a valid welfare indicator. Method and Results The play behaviour of seven litters of large white/Landrace × Hampshire piglets farrowed in the PiGsafe free farrowing system was observed over the 4 week pre-weaning period. Data was transformed for normalisation and analysed using Pearson correlations, t-tests and ANOVA where appropriate. Total play differed significantly between litters (F=27.24, P<0.001, df=6) and was not associated with overall activity levels (R^2=0.0009, P>0.05, df=6). Litter size, sex, birth weight and ponderal index were not correlated with play levels but mean growth over the pre-weaning period correlated strongly with total play (R^2=0.885, P>0.01, df=6). The overall distribution of the broad play categories of locomotor, object and social play were not different between litters (F=2.24, F=1.29, F=2.21, df=6, P>0.05, df=6). However, litters which displayed higher total play attributed a higher proportion of their locomotor play to elements 'run' and 'flop' (R^2=0.618, P>0.05 and R^2=0.912, P>0.001, df=6) while those litters showing lower total play levels attributed higher proportion of play to non-injurious fighting (R^2=-0.557, P>0.05, df=6). Males displayed higher proportional distribution of pushing behaviours regardless of the total play level of the litter (F=6.05, P<0.05, df=67). Males also performed more play initiation events (t=2.27, P<0.05, df=67) than females while the levels of play rejection was not different between sexes (t=1.85, P>0.05, df=67). These results suggest the use of play as a proxy for good welfare will only be possible when we can begin to understand individual differences in expression of play behaviour and their implications.

Faecal glucocorticoid metabolites in African Penguin: biological validation of an enzyme immunoassay

L. Ozella[1], L. Anfossi[2], F. Di Nardo[2], L. Favaro[1], D. Sanchez[3] and D. Pessani[1]
[1]University of Turin, Life Sciences and Systems Biology, Via Accademia Albertina 13, 10123 Torino, Italy, [2]University of Turin, Chemistry, Via Pietro Giuria 5, 10125, Italy, [3]ZOOM Torino, Strada Piscina, 10040 Cumiana (TO), Italy; laura.ozella@unito.it

Animals in zoos are subject to a variety of physical, ecological, and social limitations. Physiological responses to stress in captive animals can be used to evaluate their welfare. Previous studies demonstrated that exposure to stress results in an increased secretion of glucocorticoids (GCs) from the adrenal cortex. In particular, corticosterone is the main avian GC. Measurement of GCs metabolites in faeces has become a well-established method for the non-invasive evaluation of the adrenocortical activity. The biological validation is a crucial step to demonstrate that a method can detect changes in adrenocortical activity and thus providing reliable results. The aim of this study was to validate a method for non-invasively measuring GCs in faeces of the African penguin (*Spheniscus demersus*). Samples were collected from a colony in a zoological institution (ZOOM Torino, Italy) accredited by the European Association of Zoos and Aquaria. To perform the biological validation, we used a known stressful event: the capture and immobilisation of the animals for veterinary treatments. Faecal samples were collected from four adult males and five adult females before and after the stress and stored at -20 °C. After the extraction of metabolites with methanol, samples were analysed by an expressly developed enzyme immunoassay (EIA) based on the use of antibodies against corticosterone. The results showed that the peak of secretion of GC metabolites occurred 7 to 10 hours after the stressful event, both in males and females. To verify the reliability of our methodology, we compared the results obtained with those from an EIA for tetrahydrocorticosterone methodology already validated to measure GC metabolites in faeces of the Adelia penguin (*Pygoscelis adeliae*) and the Wilson's storm petrel (*Oceanites oceanicus*). The biological validation suggest that the developed EIA for corticosterone can be a useful tool for non-invasively measuring GC metabolites in faeces of the African penguin. Moreover, our results are in agreement with those obtained through the EIA for tetrahydrocorticosterone method. Therefore, we propose to exploit this methodology to evaluate sources of stress such as abundance of visitors, noise level, and the environmental constrains. Moreover, the ability to non-invasively monitor adrenocortical activity in African Penguins is important for management strategies since prolonged periods of elevated GC concentrations interfere with several physiological processes, including immune and reproductive function.

Qualitative behavioural expression of sows changes during the first 60 minutes post-mixing

T. Clarke, J.R. Pluske, D.W. Miller, T. Collins, A.L. Barnes and P.A. Fleming
Murdoch University, School of Veterinary and Life Science, South Street, Murdoch, Perth Western Australia 6150, Australia; t.clarke@murdoch.edu.au

Most welfare studies involving sows and group housing focus on the negative aspects of sow behaviour at the time of mixing, with few identifying other behavioural responses. We studied behaviour of mixed-parity sows following mixing into ten groups (each n=10) in retrofitted pens each with free-access feeding stalls (approximatly1.8 m²/sow) at 5 days post-mating. We filmed sows at the group level over the first 90 minutes and analysed behaviour using both quantitative and qualitative methods. For time budgets, the number of pigs performing each of six behavioural categories (lying, sitting, walking, standing, investigating and aggression) was recorded for each clip. Qualitative behaviour was assessed in terms of behavioural expression using Qualitative Behavioural Assessment (QBA). For QBA, 18 observers scored pigs in video clips at 10 and 60 minutes post mixing using the free choice profiling methodology. A generalised Procrustes analysis (GPA) was used to analysis the observer's scores. A comparison between the quantitative behavioural profile and the QBA assessments were investigated using a correlation matrix. The most marked differences (time budgets) were evident at two time points: 10 and 60 minutes post mixing on GPA dimension 1 (P=0.029) and GPA dimension 2 (P=0.001). Using QBA, sows were described as more active, aggressive, tense, interested and curious 10 minutes post-mixing. Sows were described as more bored, calm, relaxed, lazy and tired 60 minutes post-mixing. A correlation between quantitative and qualitative showed no significant relationships. This suggests that quantitative scoring may not be sensitive enough to reflect the dynamics of sow behaviour on a group level. However when plotted against each other the scores indicated that sows were considered more calm, content, quiet, lazy and sleepy when they were lying. Active behaviours (walking, investigating and aggression) were associated with terms such as active, curious, interested, aggravated and aggressive. Thus, QBA identified behavioural expression of sows changed over the first 60 minutes immediately following mixing. This is an important consideration when deciding the point in time at which welfare measurements are to be taken.

Cortisol and testosterone concentrations in boar bristles in relation to season and sperm viability

D. Bucci, F. Scorrano, E. Nannoni, N. Govoni and M.L. Bacci
University of Bologna, Department of Veterinary Medical Sciences, Via Tolara di Sopra 50, 40064 Ozzano Emilia (BO), Italy; eleonora.nannoni2@unibo.it

Boar welfare is a key factor in swine semen production and laboratory use. Hair analysis has been proposed as a minimally-invasive technique able of providing information on the stress response over the medium- to long-term period. Hormonal fluctuation in bristle could be a valid tool to study pig stress and metabolic responses. The aim of the present work was to determine cortisol (CORT) and testosterone (TEST) concentrations in hair, in relation to temperature and sperm viability. Two Italian Large White boars, of proven fertility, aged 6 and 9 years were used for the study; they were housed in 6 m² wide boxes, under artificial illumination and with controlled air temperature. Hair was collected monthly, semen was collected once every two weeks, environmental temperature was determined 6 times a day and mean temperature (Tmed), maximum temperature (Tmax) and minimum temperature (Tmin) were recorded every two weeks. CORT and TEST hair concentrations were measured, after one water and two isopropanol washings and methanol extraction from the matrix, by a validated RIA technique; sperm viability was determined by fluorescent probes (Propidium iodide to stain dead cells' nucleus in red and Sybr green 14 to stain live cells' nucleus in green). An ANOVA for repeated measures was used to analyse CORT, TEST and sperm viability changes during the different months; Pearson's correlation test was used to evidence correlation between CORT, TEST, viability and temperature. There is no difference in CORT, TEST and sperm viability during the different months of the year; CORT is negatively correlated to Tmin (R=-0.33, P<0.05), while only a tendency to negative correlation could be registered with Tmed and Tmax. No other significant correlation was registered. For the first time we showed the applicability of RIA for hair CORT and TEST determination in boar; the technique could be useful for large scale assessment in semen production farms as it is not invasive, relatively easy and fast. The controlled environmental condition in which the animals are kept may be the main reason of the low variation in CORT and TEST. The negative correlation between CORT and Tmin, is not surprising and could be due to the activation of endogenous heat-production systems at low environmental temperatures. On the contrary, the environmental control of high T has been quite effective in avoiding negative stressful effects on boar. In conclusion we gave some insights on new parameters that could be studied on a larger scale and could furnish some information on boar welfare and/or stress in production farms in which the environment is not as strictly controlled as in our stable.

Poster 50

A tool for transporters for self-monitoring the quality of animal transport

B. Mounaix[1], A. De Boyer Des Roches[2,3], L. Mirabito[1] and V. David[1]

[1]Institut de l'Elevage, Service Santé et Bien-être des Ruminants, 149 rue de Bercy, F-75595 Paris, France, [2]INRA, UMR1213 Herbivores, UMR1213 Herbivores, 63122 Saint-Genès-Champanelle, France, [3]Université de Lyon, VetAgro Sup, UMR1213 Herbivores, Campus Vétérinaire de Lyon 1 avenue Bourgelat, 69280 Marcy L'Etoile, France; beatrice.mounaix@idele.fr

Following the French 'Rencontres Animal et Société' in 2008 all stakeholders involved in the transport of livestock have concluded on the need to develop new tools to improve the quality of transport for the purposes of animal protection. To the request of professionals, the aim of this study was to elaborate a protocol for self-monitoring the welfare of transported animals (cattle and sheep) and the quality of equipments and management during transport. Several animal-based measures were selected based on a scientific review. At the same time, a 'hazard analysis' was carried out to select the most relevant factors, i.e. the equipment and the management of animals during loading/unloading that may impair the welfare. The feasibility of the selected measures and the inter-observer reproducibility was tested in various transport conditions: livestock markets, slaughterhouses and assembling centres (Kappa index >0.4 for qualitative measures; $r^2>0.7$ for quantitative measures; n=640 cattle; n=670 sheep). Based on these results, species specific self-monitoring marking grids have been elaborated for transporters of young calves, for cattle and for sheep to be used at the end of the transport to back-identify problems and needs for improvement. These grids were then tested for feasibility by transporters (266 unloading situations; n=809 cattle; n=660 sheep). They complement and refer to the 2009 national handbook for transporters on (EC) N° 1/2005 Regulation implementation, and to several regional leaflets for good practices and technical recommendations in loading and unloading cattle and sheep. This project has been funded and elaborated with French professionals involved in the transport of livestock to co-develop a tool adapted to their working conditions, support them in improving the quality of the transportation of farm animals and involve them practically in welfare improvement strategy.

Exsanguination blood lactate as an indicator of pre-slaughter welfare in finishing pigs

P. Brandt[1,2], T. Rousing[2], M.S. Herskin[2] and M.D. Aaslyng[1]
[1]Danish Meat Research Institute, Maglegaardsvej 2, 4000 Roskilde, Denmark, [2]Aarhus University, Department of Animal Science, Blichers Allé 20, 8830 Tjele, Denmark; pbt@teknologisk.dk

Pre-slaughter handling constitutes novel and potentially stressful experiences for pigs. The concentration of lactate in the exsanguination blood might indicate pre-slaughter stress in finishing pigs. We investigated the relationship between selected ante mortem observations and the plasma concentration of lactate (P-LAC) in 80 pigs from four herds as part of an evaluation of welfare in finishing pigs at commercial abattoirs. At the abattoir, behavioural and handling observations were carried out in the race to the stunning chamber and included slipping, falling, being moved by automatic gates and lifting by other pigs. At sticking, a blood sample was collected for analysis of P-LAC. Behavioural and handling measurements were recorded using one-zero sampling and summarized to scores 0 (no events observed), 1 (1 observed) and 2 (2 or 3 events observed). A significant relationship between the scores for behaviour and handling in the race and P-LAC was found (P=0.008). Thus, P-LAC might be an indicator of welfare of pigs at the abattoir. If P-LAC is to be used as an on-site indicator of pre-slaughter welfare at abattoirs access to a fast analysis is necessary. Comparison of the analysis of the concentration of lactate in whole blood (W-LAC) and in plasma using 107 blood samples from sticking blood at abattoirs, showed a correlation of 0.86 suggesting that W-LAC can replace the more elaborate plasma analyses. Our studies show that there is a relationship between handling and behaviour prior to slaughter and P-LAC, thus, P-LAC may indicate pre-slaughter welfare in finishing pigs. Further, our studies show that the faster whole blood analysis is applicable. More research is needed to develop an on-site analysis of W-LAC and to determine the acceptable level of W-LAC.

Poster 52

Haematologic and serum biochemical parameters in working horses

T.A. Tadich and R. Lanas

Facultad de Ciencias Veterinarias y Pecuarias, Universidad de Chile, Santa Rosa 11735, La Pintana, 8820000, Chile; tamaratadich@u.uchile.cl

Reference haematological and biochemical values are used to assess the health state of animals, but changes in baseline values of working equids could provide information on how they are coping with extreme conditions (lack of proper nutritional resources, climate, overwork). Most working equines work with some health condition, such as lameness, injuries, parasitism, or poor body condition scores; all affecting their welfare. This provides a difficulty when examining these horses and comparing their blood work with international reference values, usually obtained from Thoroughbreds. The aim of this study was to obtain haematological and biochemical values from a group of urban draught horses in Chile and compare them with national and international reference values for equines in order to assess their welfare state. Blood samples were obtained from 150 urban draught horses, mares, geldings and stallions, all over 1.5 years of age, from the cities of Santiago, Talca and Valdivia. All horses included were selected according to the criteria proposed by Pritchard *et al*. A jugular venous blood sample was obtained. Blood was transferred into tubes containing EDTA, Heparin and with no anticoagulant. All tubes were sent to the laboratories in order to obtain complete haematology and serum biochemistry. Descriptive statistics was used in order to compare our mean values with reference values belonging to Knottenbelt for horses from developed countries, and Wittwer for Chilean horses. Our main results show that all horses presented at least one haematological or biochemical value altered when compared to both, international and national, reference values. Haematological results showed RBC counts below reference value in 57% according to national reference (6.6-11.4 × 1012 /l), with a mean value of 6.57×1,012 /l. Haemoglobin was below national reference value in 45%. The haematocrit was below both national and international reference values in 37%, and the Neutrophils: Lymphocyte ratio (N:L) was above reference in 13%. The N:L ratio has been indicated as a better indicator of stress in animals than cortisol, making it a useful tool when assessing welfare. In relation to biochemical variables, 66% of horses showed selenium deficiency, and 30% had high CK values according to national reference values (40-280 UI/l) and all of the according to international reference values (<50 UI/l), CK and AST are of importance when assessing muscular damage, fibrinogen was above national reference values (1-5 g/dl) in 27%, and in a higher percentage when using international reference values. Urban draught horses in Chile work under altered physiology, which could indicate subclinical health issues or adaptation to their context.

Evaluation welfare of dairy cows conducted in tied up mode using Welfare Quality® protocol

A. Benatallah[1], F. Ghozlane[2] and M. Marie[3]
[1]Higher national veterinary school of Algiers, Veterinary, 16000, Hacen Badi El Harrach, Algiers, 16000, Algeria, [2]Higher National Agronomy School of Algiers, Algeria, animal husbandry department, 16000, belfort El Harrach, Algiers, 16000, Algeria, [3]INRA-ASTER-Mirecourt and University of Loraine ENSAIA, veterinary, 662 Av. Louis Buffet, 88500 Mirecourt, France and, 2 Avenue de la Forêt de Haye, TSA Vandœuvre cedex, 88500, France; safoutou@yahoo.fr

The animal welfare is complex due to the multiplicity of its components. This concept has emerged in societies as critical interrogation on the living conditions of animals in intensive system. The negative influence of this system is accompanied by a rise of concerns of animal welfare (legislative, ethical, sociological, philosophical and scientific). The scientific study of animal welfare has grown rapidly in recent years and several evaluation methods have been developed. Among them, the welfare quality protocol (2009) from the welfare quality project that includes not only several aspects of animal welfare (hunger, thirst, health, animal behaviour...) but allowed also to answer at several concerns of farmers and serve them as a tool for diagnosis and advice. Our study aimed to identify negative and variable aspects of dairy cow welfare in farms conducted in tied up mode and to determine which farm characteristics were associated with impaired welfare. We assessed welfare using the Welfare Quality® protocol in 53 dairy farms in the province of Algiers. Within each farm, scores that express the degree of farm compliance with 11 welfare criteria were calculated (absence of hunger, thirst, diseases, alteration of integument, comfort around resting, and normal behaviour...). We used linear models (analysis of variance used unbalanced data) to assess the association between the characteristics of farm (herd size and breed) and scores of welfare ($P<0.001$). The results showed a degraded level of welfare with 53 farms downgraded (less than or equal to 10 score on a principle or less than or equal to 20 on at least two principles). The most degraded scores related to principles: good feeding (5.2 ± 1.1); good housing (20.2 ± 2.00); good health (41 ± 9.7) and all criteria excepted: absence of pain caused by management procedures (100 ± 0), emotion (85 ± 14.7) and agonist behaviour (67 ± 7.6). Variability's were found between large and small farm ($P<0.001$). This study helped to determine the level of farm welfare surveyed conducted in tied up mode and identify the risk factors that contributed to its degradation in order to make improvements to their levels.

Carcass skin lesions in boars reflect their aggressiveness on farm

D.L. Teixeira, N. Van Staaveren and L.A. Boyle
TEAGASC, Pig Development Department, Moorepark, Fermoy Co Cork, Ireland;
dayane.teixeira@teagasc.ie

The aim was to investigate if higher incidences of aggressive and mounting behaviours in boars are associated with higher skin lesion scored on farm and at abattoir compared to gilts. A total of 141 boars and gilts were housed in single-sex groups (n=5 of each, mean group size was 14) at finisher house (0.73 m²/pig). Data were collected throughout 2 weeks prior slaughter (Day -14) and 2 h later (Day -1). Posture (lying, sitting and standing), aggression (head knock and fight) and mounting behaviours were recorded in 3 periods (8-10, 11-13, 14-16 h) on days -13, -9, -7 and -2. On days -14 and -1 pigs were weighted and skin lesions scored according to severity. Pigs were mixed prior transport and slaughtered in an abattoir distanced 99 km from the farm. Carcass cold weights were obtained from abattoir data. At chill room, skin lesions were assessed as per Welfare Quality® protocol (WQ) and bruises were counted and classified as fighting-, mounting- and handling-type. Data were analysed in SAS (performance and behaviour: PROC MIXED; scores: PROC NPAR1WAY; association between variables: PROC CORR). Boars were slaughtered heavier than gilts (100.7 vs 99.0 kg; s.e.m. 0.59; P<0.05) but there was no effect of gender on carcass cold weight (77.3±0.4 kg). Boars performed more aggressive behaviours than gilts (1.76 vs 1.03 aggression/pig; s.e.m. 0.22; P<0.05), as well as more mounts (0.40 vs 0.005 mounts/pig; s.e.m. 0.02; P<0.05). Lying active (5.0±0.8%), lying inactive (64.9±2.8%) and sitting (6.8±0.6%) behaviours were similar in both genders, as well as time standing at periods 1 (31.6±2.9%) and 3 (20.5±2.9%). But boars spent more time standing at period 2 than gilts (21.8 vs 13.7%; s.e.m. 2.86; P<0.05). On Day -14, total skin lesion scores were similar in both genders, but on day -1 boars had higher scores than gilts (11.17 vs 8.2; s.e.m. 0.95; P<0.05). From carcasses, boars had higher total skin lesion score (1.88 vs 1.3; s.e.m. 0.1; P<0.05) and more fighting-type bruises (4.45 vs 2.25; s.e.m. 0.35; P<0.05) than gilts. There was no association between aggressive behaviour and skin lesion scored on farm on Day -1 (actor: r=0.004, P>0.05; recipient: r=0.022, P>0.05) but there were positive correlations between aggressive behaviour and skin lesion score on carcasses (actor: r=0.383, P<0.0001; recipient: r=0.294, P<0.001, respectively) and fighting-type bruises (actor: r=0.442, P<0.0001; recipient: r=0.297, P<0.001, respectively). There was no association between skin lesion scores recorded on farm and at abattoir (r=-0.093, P>0.05). Carcass skin lesions scored at abattoir were more sensitive to detect the level of injury and poor welfare caused by the aggressiveness of boars.

Potential pain indicators in tail docked lambs

J. Marchewka[1], I. Beltrán De Heredia[1], R. Ruiz[1], J. Arranz[1] and I. Estévez[1,2]
[1]*Neiker-Tecnalia, Animal Production, Campus Agroalimentario de Arkaute, 01080 Vitoria-Gasteiz, Spain,* [2]*IKERBASQUE Basque Foundation for Science, Alameda Urquijo 36-5, 48011 Bilbao, Spain; jmarchewka@neiker.net*

Pain reactions in ungulates are hidden, possibly as an anti-predator response to diminish predation risk and therefore are difficult to detect. However, exacerbation of the pain response might be obtained in situations that elicit fear reactions such as separation from flock mates, facilitating the identification of behavioural responses associated with pain. The aim of this study was to try to identify potential behavioural pain indicators by analysing the behavioural responses of tail docked lambs on which the procedure was conducted with or without pain control measures. Twenty four female lambs were divided in four equal groups, each assigned to one of the treatments: tail docking (TD) with rubber ring, TD with anaesthesia (TDA), TD with anaesthesia and analgesia (TDA+A) and control (C) handled but without TD or pain relieve application. Individual lambs underwent 2.5 minute open field test the day prior-, day of the procedure, and 1, 3 and 5 days post-procedure, to determine the effect of treatment on lambs responses and their post-procedure evolution. Behavioural responses of the lambs were recorded during the test with the Chickitizer software and included: rest, run, stand, walk and explore. Resting was close to zero and therefore disregarded. Videos of the tests were also recorded and were used to extract the frequency and durations of climbing and look over the walls (CLW). In addition, blood samples were taken on days -2, 0, 2, 4 and 6 for cortisol evaluation (Cort). We built a GLM model including treatment and day as fixed and lamb as a random effect, with day as repeated measure. Treatment effects were detected for standing, walking and CLW ($P<0.05$) but not for exploration and running ($P>0.05$).TD lambs stood less than all others and walked less than C lambs ($P<0.05$). Total duration of CLW ($P=0.0082$) was highest for C lambs as compared to all others, while total CLW frequency ($P=0.02$) was highest for TD lambs. When distinguishing longer exploratory climbs (EC), in which lambs stood observing the adjacent pen, from the abrupt ones (AC) lasting up to 2 s, a treatment effect was found ($P=0.03$) with TD lambs showing a higher frequency of AC as compared to TDA and TDA+A ($P<0.05$) and was close to significant when compared to C lambs ($P=0.07$). A day effect was found for all behaviours although results are difficult to interpret. In addition we found significant effect of treatment on the Cort levels ($P=0.03$) with C lambs showing lower levels as compared to TDA. These results suggest that under the conditions of isolation of the current study, animals undergoing potentially more severe pain stood and walk less while showed higher AC frequency.

Development of a protocol to assess the animal welfare of sheep

I. Beltrán De Heredia[1], S. Richmond[2], F. Wemelsfelder[2], R. Ruiz[1], J. Arranz[1], E. Canali[3] and C.M. Dwyer[2]

[1]*Neiker Tecnalia, Animal Production, Vitoria-Gasteiz, 01080, Spain,* [2]*SRUC, Animal Veterinaty Science, Edinburgh- Scotland, EH25 9RG, United Kingdom,* [3]*University of Milan, Animal Science, Via Festa del Perdono, 20122 Milano, Italy; ibeltran@neiker.net*

The diversity in practices and farming conditions of sheep observed in the European systems influence the productivity, behaviour and welfare status of the animals. Especially in the case of pasture based farming systems, reproduction, feeding practices and even genetic potential of breeds, have been tailored to the seasonal availability of natural resources. However, animals may usually have to remain indoors kept during certain periods of the year due to harsh winter conditions, lack of pasture growth, or to better manage crucial productive periods (lambing, lactation, etc.). Therefore, an animal-based protocol to assess the welfare status of sheep should be developed considering this diversity. A two-step methodology has been developed within the EU funded AWIN project, with the objective of screening farms in stage 1 according to welfare issues, and then trigger a more detailed second stage if welfare concerns are detected. Basically, every group of sheep across the different housing systems available on the farm should be sampled to represent all the physiological status present at the evaluation. Animal-based assessments include: (1) QBA (Qualitative Behavioural Assessment); (2) Quantitative Behavioural assessments, such as the proportion of ewes standing, lying, panting, scratching and vigilant, taken at a sufficient distance so that the sheep are undisturbed; (3) Physical status, to assess Coat cleanliness, Fleece length and quality, tail length and incidence of lameness; (4) Fear assessment, through the evaluation of the flight distance of sheep to the approach of humans and startle responses. At stage 2, a sample of sheep are evaluated individually: Breed, age and reproductive status of each ewe, together with time of day and weather conditions is recorded, and a detailed physical assessment made of: teeth status; colour of mucosa; eyes; nasal discharge; ears; horns; respiration; skin lesions; fleece length and quality; coat cleanliness; evidence of lesions or callus in legs; body condition score; conditions of the udder and teats, and mastitis; dag score; tail length and lameness. Some basic information is necessary regarding environmental conditions (water availability, shelter, food or pasture availability, landscape, topography, etc.) to supplement animal-based assessments. This prototype is being tested in commercial flocks representative of different farming conditions and objectives. The current status of the prototype and partial results of the test will be presented.

The development of the fox protocol: the second refinement phase

J. Mononen[1,2], T. Koistinen[2], H. Huuki[2] and L. Ahola[2]
[1]*MTT Agrifood Research Finland, Animal Production Research, Halolantie 31A, 71750 Maaninka, Finland,* [2]*University of Eastern Finland, Department of Biology, P.O. Box 1627, 70211 Kuopio, Finland; jaakko.mononen@uef.fi*

The development of fox and mink on-farm welfare assessment protocols started in 2009, and has based on the framework provided by the Welfare Quality® (WQ) protocols and project. The implementation of the fox protocol with 15 output-based and 10 input-based measures is being tested on 84 Finnish fox farms. Here we use our practical experiences from this testing phase and preliminary calculation of the scores to discuss the putative needs to refine the fox protocol at the measure and scoring levels. 'Use and type of neck tongs' measure turned out to be difficult to assess in a plausible way and has already been removed from the protocol. Some of the measures, e.g. 'Killing method', are poor in differentiating farms, at least in Finland. 'Space available for moving' needs to be redefined to include the size of the animals. 'Opportunity to use enrichment' is an example of a measure that would benefit from simplifying the scoring system. 'Impaired mouth and teeth health' and 'Urinary tract infections' are rare, and they can be included in the 'Obviously sick fox' measure. 'Temperament test' (poor validity and feasibility) and 'Feeding test' (poor feasibility) could be removed and replaced with subjective temperament scoring. Temperament and 'Stereotypical behaviour' could be measured during taking the health measures, which would shorten the time needed for an assessment visit. Simplifying and shortening the sampling procedure would also facilitate taking all measures in all of the three production periods: winter, summer and autumn. Many of the changes suggested above for the measures would have effects on the scoring system. If the scoring system is changed to the next version of the fox protocol, at the same time even bigger structural changes could be considered. Firstly, pain and distress could be included into one criterion to form 'Absence of pain or distress induced by management procedures', since there is little evidence of pain being caused by any of the fox management procedures, but assumingly many procedures cause distress. Secondly, the relative importance of the various criteria (based on WQ) within the 'Appropriate behaviour' principle should be considered, because of the lack of plausible measures for the 'Positive emotional state' criterion. Thirdly, it could be better to calculate first principle scores separately for each production period and aggregate them only after that into an overall evaluation of the farms. This would make it easier to use the assessment results for advisory purposes.

Development of the WelFur on-farm welfare assessment protocol for the Finnraccoon

T. Koistinen[1], H. Huuki[1,2], J. Mononen[1,2] and L. Ahola[1]
*[1]University of Eastern Finland, Department of Biology, P.O. Box 1627, 70211 Kuopio, Finland,
[2]MTT Agrifood Research Finland, Animal Production Research, Halolantie 31A, 71750 Maaninka,
Finland; tarja.koistinen@uef.fi*

WelFur on-farm welfare assessment protocols were developed for foxes and mink in the European WelFur project. Also the Finnraccoon (*Nyctereutes procyonoides ussuriensis*) is an important fur animal species in Finland, with an annual production of 160,000 pelts. Here we describe the development of the WelFur on-farm welfare assessment protocol for the Finnraccoon. Finnraccoons are housed under similar circumstances as foxes, and the management resembles much that of foxes. Therefore, the WelFur fox protocol was used as a starting point. However, since some physiological and behavioural characteristics of Finnraccoons differ from those of foxes, all measures, e.g. Ocular inflammation and Availability of platform, could not be adopted as such; instead, the special characters of Finnraccoons, e.g. their tendency to intermittent hibernation in winter, were included into the protocol. The first proposal of 29 measures to be included in the protocol, made by the Finnish researchers involved in the development of WelFur protocols, was tested on 12 Finnraccoon farms during the three production periods recognised earlier for foxes and mink: (1) breeding animals in winter; (2) breeding animals during cub nursing period in summer; (3) breeding animals and growing juveniles in autumn. An expert panel consisting of six ethologists, three veterinarians, two fur farmers and one fur production teacher discussed the initial proposal. Based on the discussions, 12 animal-based and 11 input-based measures were included into the final protocol. Of all measures, eighteen are measured in all three, two in two, and three in one of the production periods. The scoring system was developed with the aid of the same expert panel. Similarly to the WelFur fox and mink protocols, each expert scored the possible outcomes of the measures and aggregated the measures within each welfare criterion. Unlike for the WelFur fox and mink protocols, the same experts aggregated also the criteria scores into the principle scores. The calculation model was kept more robust than in the WelFur fox and mink protocols, e.g. regression models were applied instead of the l-spline functions. The synthesis of the four principles scores into an overall assessment was based on the Welfare Quality® logic, used also in the WelFur fox and mink protocols, but unlike in these protocols the overall farm-level assessments were made separately for each production period. The experiences from developing and testing of implementation of the WelFur fox and mink protocols were extremely useful in the development of the Finnraccoon protocol.

Assessing cow-calf or suckler cow herd welfare: a pilot study on 10 California ranches

G.E. Simon, B.R. Hoar and C.B. Tucker

University of California, Davis, One Shields Ave, Davis, CA 95616, USA; gsimon@ucdavis.edu

There is growing concern about the welfare of animals used for food production in the United State. Certification programs are one way ranches can assess and communicate information about animal welfare to the public. To date, there is no assessment tool addressing welfare specifically on cow-calf operations (or suckler cow herds) in the U.S. The objective of this study was to develop and test the feasibility of an animal welfare assessment tool for cow-calf operations. Based on similar welfare programs in the beef and dairy industries, an assessment tool was developed and tested on 10 California ranches. The assessment included a ranch visit to observe cattle health and stockperson handling during routine procedures and a questionnaire to gather herd management information collected in a survey (n=5) or interview (n=5) format. Six hundred and eighty six Black Angus, Red Angus, Angus cross and Simmental cattle (35% cows/heifers, 65% calves) were observed being processed in the chutes. Procedures ranged in invasiveness from branding to recording body weight and the duration of restraint was between 0.5 minutes/animal (recording body weight) to 6.4 minutes/animal (freeze branding). In the chute-side observations, stockperson handling techniques (e.g. electric prod used on 0 to 95% of animals observed in a given herd) and cattle response to handling (e.g. 0 to 69% of given herd exited chute at a run) varied considerably between ranches. Health-related issues (e.g. 0 to 13% of given herd with low body condition) were rare and only seen on specific ranches. Other problems, such as bloated rumens and broken tail heads were never observed. Information collected in the questionnaire was best obtained in an interview format. Similar to other welfare assessment schemes in the U.S., 50% of the questionnaire inquired if the ranch maintained written management protocols for practices such as castration and dehorning. Most ranches did not maintain such documents, suggesting that animal-based measures and verbal descriptions of management practices may be a more appropriate method of capturing the desired information. The results of this pilot study have identified areas where welfare may be compromised on U.S. cow-calf operations (i.e. cattle handling) and will inform future design of assessment tools. Expanding the study to include additional ranches will allow for epidemiological evaluation of risk factors and to inform or serve as the basis for certification programs.

Poster 60

Assessing seasonal variation in welfare indicators in extensively managed sheep

C.M. Dwyer[1], S. Richmond[1], F. Wemelsfelder[1], I. Beltran[2] and R. Ruiz[2]
[1]*SRUC, Animal and Veterinary Sciences, King's Buildings, West Mains Road, Edinburgh, EH9 1BN, United Kingdom,* [2]*Neiker Tecnalia, Campus Agroalimentario de Arkaute, 01080. Vitoria-Gasteiz, Spain; cathy.dwyer@sruc.ac.uk*

Welfare assessment in extensive environments presents particular challenges, including difficulty in observing the animals in an undisturbed state, and seasonal variation in climate and food availability. Additionally, sheep breed seasonally, such that animal can be in different stages of reproduction at different times of the year. To define the seasonal variation in welfare indicators in sheep, 100 sentinel Scottish Blackface ewes were marked within a larger flock of approximately 700 ewes and observed for two consecutive years. These animals lived outdoors all year round, usually grazing unimproved hill pastures, and were managed under commercial conditions. On 5 occasions during each year, at planned management gathers, the ewes were identified and scored for the welfare indicator protocol previously developed by the EU-FP7 'AWIN' project. Indicators assessed measures of health, body condition and physical appearance, and included blood counts and faecal assessments of parasite burdens. In each year, independent of welfare indicator assessment, a subset of ewes was drafted out of the flock by the shepherd for management reasons (usually health or productivity). The effects of time of year and likelihood for leaving the flock were assessed for each indicator using linear regression models or $\chi2$ tests for binary data (present/absent). Time of observation had a significant effect on body condition score (BCS), tooth loss, nasal discharge, and anaemia scores (P<0.001 for all), with variation within and between years. In both years ewes in mid lactation and at weaning had poorer health and BCS than at other times, and year 2 scores were lower than year 1 (e.g. Mean BCS: Year 1: mid-pregnancy=2.78, mid-lactation=2.42; Year 2: mid-pregnancy=2.37, mid-lactation=2.00, s.e.d.=0.19). Ewes that were removed from the flock had significantly lower BCS and greater tooth loss scores than retained ewes. Between 8 and 12% of ewes were lame at each assessment but lameness score did not show seasonal variation nor did it differ between ewes that were removed or retained in the flock. However, there was a significant interaction between retention and time of year (P=0.006) with retained ewes having greater lameness at mid lactation and weaning whereas no pattern of seasonal variation in lameness was present in the draft ewes. The data show that, within a single farm with consistent management, significant seasonal variation occurs which needs to be adjusted for when conducting welfare assessments. In general, ewes in mid and late lactation were in poorer health than at other times.

Poster 61

Quantitative sensory testing for assessing wound pain and anaesthesia in livestock

S. Lomax, P.A. Windsor and P. White
University of Sydney, Faculty of Veterinary Science, RMC Gunn Bldg B19, University of Sydney, NSW 2006, Australia; sabrina.lomax@sydney.edu.au

Quantitative sensory testing (QST) is a validated technique that we have adapted for our research to record the evolution and distribution of pain from mulesing, tail docking, castration and dehorning wounds, and asses the efficacy of analgesic treatments. In our studies we have used Von Frey (VF) monofilaments and pressure algometry to assess wound pain and anaesthesia. VF monofilaments are calibrated to bend at a predetermined pressure in order to provide repeatable pain stimulation of sites on the wound and surrounding skin. Responses are scored using a customized numerical rating scale (NRS), by monitoring induced involuntary motor reflexes in the rump and head. Pressure algometry is used to measure pain by means of pressure transduction. Pain threshold is recorded as maximum pressure (g/f) exerted before animal response and withdrawal. We performed QST using mechanical stimulation with both VF monofilaments and algometry to evaluate hypersensitivity after minor surgery in lambs (castration, tail docking, mulesing) and calves (castration and dehorning). A strong decrease in mechanical thresholds proximal and distal to the wound was indicative of pain. Additionally, these techniques allowed us to assess the efficacy of a novel topical anaesthetic (TA) treatment for wounds. VF stimulation of mulesing, castration and tail docking wounds in lambs indicated significant hypersensitivity of the wound site within 3 minutes of surgery, with pain sensitivity increasing up to 8 h. This was significantly higher than lambs treated with TA ($P<0.01$). Similar results were seen in calves following VF stimulation of castration and dehorning wounds in calves. Results from pressure algometry have supported our VF results, adding more objective and quantifiable data. TA treated calves exhibited significantly greater pain threshold of the castration wound (559.2 ± 14.3 g) than untreated calves (446.0 ± 18.9 and $P<0.001$). In a recent study we found that TA treated calves also had significantly greater pressure threshold (3.89 kg/f) of the dehorning wound than conventionally dehorned calves (1.77 kg/f, $P<0.001$). We are currently using this technique to measure skin sensitivity in piglets and to assess the efficacy of cold induced anaesthesia. This presentation will use our research findings to demonstrate how QST provides us with an important tool for understanding the generation and development of wound pain, for clinically assessing and quantifying animal pain, and for assessing the efficacy of anaesthetic and analgesic options.

Pedometry for the assessment of pain in 6-month old *Bos indicus* calves following surgical castration

M. Laurence, T. Hyndman, T. Collins and G. Musk

Murdoch University, College of Veterinary Medicine, South Street, Murdoch, 6150, Australia; m.laurence@murdoch.edu.au

In Australia, analgesia is not usually administered to cattle undergoing surgical castration on the farm. Furthermore, the pain associated with this husbandry practice has not been adequately characterised. To investigate analgesic strategies for this procedure, pedometry was used to assess the efficacy of lignocaine (L) and meloxicam (M) administered in the perioperative period. It was hypothesised that the calves suffering post-surgical pain after an open castration would be less active (fewer steps per day and more rest bouts) than those that either did not have surgery or those that had received analgesic drugs. The perioperative analgesics were 0.5 mg/kg subcutaneous (SC) M and 2 mg/kg of intra-testicular and SC L. 48 six-month old Brahman bull calves were randomly divided into six equal groups: no surgery control; surgical castration without analgesia; castration with M_{pre-op}; castration with $M_{post-op}$; castration with L and $M_{post-op}$; castration with L. Electronic pedometers were strapped to the left hind leg five days before surgery. Cattle were gathered in a race daily from the day the pedometers were fitted to 13 days after surgery. Activity (steps per day) and rest bout data (number of periods of recumbence) were downloaded from individual pedometers using a hand held wand. Data were analysed with a mixed effect linear model with activity or rest bout as the response variable and day and analgesic treatment as predictors (P<0.05 was considered significant). Data are expressed as deviations from the control treatment ± SE. Plasma cortisol was measured every second day. The day after surgery, calves that received pre-operative M were more active than all other castrated calves (P=0.038, 6.5±19 vs -32.9±20 steps). All castrated animals rested more than the no-surgery control group (P=0.013). Calves that received analgesia rested less than those that received no analgesia (P<0.001, 0.27±0.3 vs 0.95±0.25). There were no significant differences between treatment on other days. Calves that received pre-operative M had, on average, lower concentrations of plasma cortisol than all other castrated animals (P=0.041). Calves that received no analgesia had higher average plasma cortisol concentrations than all other castrated animals (P=0.08). Pedometry is a useful way to measure the efficacy of perioperative analgesia in six-month old castrated *Bos indicus* calves. Pain responses in cattle are difficult to measure but this methodology may be a useful tool for the assessment of acute pain in cattle.

Development of a Welfare Quality-like protocol for sheep in the Netherlands

P. Koene and J. De Jong
Wageningen Livestock Research, Animal Welfare, De Elst 1, 6708 WD Wageningen, the Netherlands; paul.koene@wur.nl

The lack of a WQ* protocol for sheep might be due to the high level of naturalness in sheep husbandry. Despite the fact that sheep are often housed outside and their welfare looks adequate many problems occur in sheep husbandry. Diseases and high lamb losses are quite common in sheep husbandry. A proper protocol is thus needed to assess the welfare and give farmers a guideline for housing, health, feeding and behaviour in sheep. The aim of this study was to create a Welfare quality-like protocol for sheep in the Netherlands. The protocol was developed using the Welfare Quality protocol for cattle and the framework created by Goddard. The WQ-like protocol was developed with special attention for the human animal relationship (HAR). The relationship between the HAR and the reproduction rate was studied as well as the connection between the HAR and other parameters of the WQ-like protocol. To test the developed protocol 6 farms in Noord-Holland (the Netherlands) were visited of which 5 were average size (<100 sheep) and 1 large farm (700 sheep). These 6 farms cooperated with testing the WQ-like protocol and the HAR test in the period between April and May 2012. This period was used as most lambs are born then and the reproduction rate was easier to measure. A relatively low amount of farms was willing to cooperate in creating the protocol. Parameters added to the protocol were access to pasture, presence of rough terrain, and health parameters like hyperkeratosis, foot rot, lamb losses and castration. The highest score to achieve for all parameters in the WQ-like protocol together is 36 points. The highest score is given to farm 3 (28 points), the lowest score to farm 5 (23 points). With respect to the HAR test, sheep that show less fear for unfamiliar humans show a lower amount of lamb losses when compared to more fearful sheep (R=-0.16, P<0.05). The less fear for humans seems to be affected in a positive way by the presence of dry lying area (R=0.43, P<0.01) and a better relationship with the stockperson (R=0.34, P<0.01). No significant relation was found between the approach/avoidance distances (HAR) and the reproduction rate of the farms. In conclusion the results show that the developed protocol discriminates between sheep farms and can be used as a guideline for creating a complete WQ* protocol for sheep. Moreover this research gives an indication on how reproduction rate is related to parameters measured in the developed protocol in sheep. Further research and more on-farm testing are needed to validate and develop a WQ* protocol that is complete.

Social networks and welfare in groups of zoo and farm animals: examples of different species

P. Koene and B. Ipema
Wageningen Livestock Research, Animal Welfare, De Elst 1, 6708 WD Wageningen, the Netherlands; paul.koene@wur.nl

Hale proposed that certain wild animal characteristics favoured domestication. Indeed, domestic animals tend to be large, non-selective feeders occupying open habitats. They are socially organized non-territorial species, typically occurring in relatively large groups in their natural environments. Such grouping of animals has costs and benefits. A strong social network is needed to maintain the group and cope with many challenges from their environment. In captivity such networks may or may not exist, dependent on actual group size and living conditions. It may be of advantage to keep animals in the groups with the social networks they are adapted for. Data on the social networks of animal species are mostly lacking and are needed to recognize the importance of a network for animal species. In addition, deviations of the social networks of animals as for instance an individual that associates increasingly less with a network may give indications of potential problems of health or welfare. Social Network Analysis (SNA) allows to characterize subgroups, transfer of information and individual actions and preferences. Social networks are currently often investigated for several reasons of which social welfare of animals is a promising one. In many cases social networks might be important for fitness, survival and welfare. We describe small social networks in horses, bears, chicken and veal calves to illustrate the increasing importance of measuring their social network for animal management. Validation of a network is shown in a group of mares and foals based on nearest neighbours with behavioural observations (allogrooming, P<0.001). SNA information is used to guide welfare-friendly removal of foals from the herd. Pitfalls and opportunities are investigated further in the network of a group of brown bears using information of solitary and territorial individuals. The social network of chicken is investigated using 1-mode and 2-mode networks related to conspecifics and the available facilities for eating, drinking, scratching and perching. SNA helped in identifying feather peckers. Finally, we show and emphasize the automatic measurement of location and/or nearest neighbours for management purposes using data on veal calves. SNA in veal calves in a pen with cubicles, lying area and feeding places showed significant daily changes in their social network, indicating instability of the social relations. We conclude that social networks are important for the welfare management of captive animal species and that automatic recording and follow-up management actions are feasible in the near future.

Iceberg indicators: fact or fiction?

C.A.E. Heath, W.J. Browne, S. Mullan and D.C.J. Main
University of Bristol, School of Veterinary Sciences, Langford House, BS40 5DU, United Kingdom; cheryl.heath@bristol.ac.uk

The theoretical concept of 'iceberg indicators' suggests that by measuring only a subset of animal-based measures it may be possible to predict the overall welfare state of an animal. To test the concept of iceberg indicators, a Welfare Quality assessment was carried out on 92 UK dairy farms which were found to have an overall Welfare Quality classification of either 'Enhanced' or 'Acceptable'. Inspection of the correlations between the animal-based measures showed no evidence in support of iceberg indicators. Next, logistic regression models were fitted using subsets of the measures, and cross-validation was used to examine how well such models predicted the overall classification. The proportion of times that the method correctly predicted the overall Welfare Quality classification increased when the aggregated score for 'Absence of prolonged thirst' was included in the model. As a single variable, 'Absence of prolonged thirst' correctly predicted the overall classification 88% of the time. A single, resource-based measure driving the classification system is, however, at odds with the conceptual underpinnings of the protocol which, instead, espouses a multidimensional, animal-based account of welfare. It is therefore suggested that the prominence of 'Absence of prolonged thirst' in this role may be better understood as an unintended consequence of the published measure aggregation system rather than as reflecting a realistic iceberg indicator.

CCM; an innovative tool to assess and control animal welfare in the pork chain

D. Oorburg[1,2], L. Heres[1] and H.A.P. Urlings[1,2]
[1]*VION Food Group, Quality Department, Boseind 10, 5281 RM Boxtel, the Netherlands,*
[2]*Wageningen University, Department of Animal Sciences, Animal Nutrition Group, De Elst 1,
6708 WD Wageningen, the Netherlands; derk.oorburg@wur.nl*

The abattoir is an interesting point to assess animal welfare in the meat production chain due to the fact that it receives animals from many farms, being all different epidemiological units. By means of continuously monitoring every delivery of animals to the abattoir (outcome based) data can be gathered to predict the level of animal welfare control prior to the arrival at the abattoir. These data are used to develop Key Performance Indicators (KPI) to control animal welfare in the animal supply chain. On a Dutch pig abattoir KPI's such as stocking density, transportation time, the number of animals with gross lesions found at the ante mortem inspection and the number of animals that need immediate attention were identified. These were used to assess the level of animal welfare during transportation to the abattoir and some relevant aspect on the farms. Large amounts of data from the abattoir were evaluated. Both the abattoir and the competent authority agreed on the relevant KPI's These KPI's are assessed continuously by comparing them to pre-set performance levels. Exceeding these levels implicates room for improvement for animal welfare in the supply chain. This information is reported to the farmer or transport company. They are demanded to take action to improve welfare control. The animal welfare performance is continuously assessed of each individual farm and haulier. With this approach we were able to control animal welfare based on outcome data in the pork supply chain in the Netherlands resulting in a better control of animal welfare. The Dutch competent authority use the same KPI's to evaluate the control of animal welfare at the slaughterhouse besides their regular audits. Using objective data to produce relevant information on the level of animal welfare control, Food Business Operators are able to continuously improve animal welfare in their supply chain.

Applicability of the transect methodology to assess on-farm turkey welfare: the i-WatchTurkey app

I. Estevez[1,2], J. Marchewka[2], V. Ferrante[3], G. Vezzoli[4] and M.M. Makagon[4]
[1]*IKERBASQUE, Basque Foundation for Science, Alameda Urquijo, 36-5, 48011 Bilbao, Spain,* [2]*Neiker Tecnalia, Animal Production, P.O Box 46, 01015 Vitoria-Gasteiz, Spain,* [3]*Università degli Studi di Milano, Via G. Celoria 10, 20133 Milan, Italy,* [4]*Purdue University, Animal Science, 125 S. Russell St., 47907 W. Lafayette, IN, USA; iestevez@neiker.net*

On-farm welfare assessment of commercial meat poultry is challenging due to the large size of the flocks. In turkey production, the size and flighty nature of the birds, further complicates the assessment. Transect walks, a new approach for general health and welfare assessment of meat poultry, recalls on the routine screening procedures of farm managers, overcomes sample size issues by evaluating virtually the entire flock, and does not require disruption of birds by catching and handling. The transect methodology requires observers to walk along standardized paths while recording incidences of birds with major welfare and behavioural problems. Our aim within the FP7 EU AWIN project was to test the viability of the transect methodology for on-farm welfare assessment of commercial turkeys. The initial study was conducted in broilers because welfare indicators have been validated in broilers, and to avoid potential constraints of the initial study related to the large size turkeys. Results of the transect walks showed high inter-observer reliability for both species. In addition, for broilers bootstrapping results revealed that a reliable house mean can be obtained by assessing only 20% of the house. Compared to traditional sampling, transect walks for broilers differed in the frequency of the incidence of common indicators such as immobility and lameness. Similarly, in turkeys we found frequency differences for most indicators between transect walks and the non-handling sampling method, but not between transect walks and frequencies obtained through individual evaluation of all the birds when funneled out of the barn during load out. The results obtained in both studies provide supporting evidence that the transect approach might be a simple, but reliable, on-farm assessment method of economically important welfare indicators. The method has the advantage that is readily acceptable and easy to implement by producers, and requires minimum training as it evaluates obvious severe welfare and behaviour problems. Based on this methodology an app for turkey health and welfare assessment (i-WatchTurkeyâ) has been designed. This app allows standardized data collection of critical welfare indicators based on the transect methodology and permit farm managers to screen the performance of their flocks over time. The main asset of this innovative solution is its simplicity of data entry, and the option to get the report immediately after assessment.

Welfur mobile, a mobile application for measurements in farms

M. Reichstadt, R. Botreau and Y. Gaudron

INRA, route de Theix, 63122 Saint Genes Champanelle, France; matthieu.reichstadt@clermont.inra.fr

Welfur is the welfare program launched by the European Fur And Breeder Association (EFBA) for fur animals. The objective of the WelFur program is to set a general certification protocol at European farm level, which will guarantee a high level of animal welfare on fur farms. WelFur is designed to be implemented directly at the farm, is based on two protocols (mink, fox). It calculates scores for 4 welfare principles (12 welfare criterions): (1) good feeding (absence of prolonged hunger, absence of prolonged thirst); (2) good housing (comfort around resting, thermal comfort, ease of movement); (3) good health (absence of injuries, absence of disease, absence of pain induced by management procedures); (4) appropriate behaviour (expression of social behaviours, expression of other behaviours, good human-animal relationship, positive emotional state) The computing part of the project is divided into two applications. The website(https://www1.clermont.inra.fr/welfur) presents the statistics of the WelFur project and allows registered users to work on the farms they assess. Each expert can import the measures collected from the farms. These data are calculated by the system using the protocol to provide scores for each criterion. These 12 scores are then computed to provide scores for each principle. Then these 4 scores are computed to give an overall score which will define the quality of the farm. The expert can access all these results in order to see whether the farm he assessed is good or not. He can also simulate changes for each measure he made, so he can point where the farm can improve. Finally he can export all the results to present them to the farmer The assessing part is made of an application used by the expert directly on the farm. Farm data are collected for the 3 periods of the annual production cycle. Because assessing a farm manually on a paper was too long and heavy, we developed a mobile application available both on windows and android tablets. Once connected, the expert can choose which specie he wants to assess. Going through the application he fills the information needed for all criterions. He can process cage by cage for animal-based and resource-based measures, shed by shed for stereotypic behaviours, and feeding capacity. The system informs the expert how many individuals are still needed to complete the visit. Once all information is collected, these raw data are transformed in Excel sheet by the program. The expert can import this Excel file on the website. He can also directly show the results to the farmer at the end of the visit. The goal of this application is to ease the work of the expert on the farm. It can be downloaded from the Welfur Project website, with a documentation for non-experts.

Artificial neural networks for revealing important information in goat contact calls

A.G. Mcelligott[1], E.F. Briefer[2] and L. Favaro[3]
[1]Queen Mary, University of London, Biological and Experimental Psychology, School of Biological and Chemical Sciences, Mile End Rd, London, E1 4NS, United Kingdom, [2]ETH Zürich, Institute of Agricultural Sciences, Universitätstrasse 2, 8092 Zürich, Switzerland, [3]University of Torino, Department of Life Sciences and Systems Biology, Via Accademia Albertina 13, 10123 Turin, Italy; a.g.mcelligott@qmul.ac.uk

Machine learning techniques are becoming an important tool for studying animal vocal communication. The goat (*Capra hircus*) is a very social species, in which vocal communication and recognition are important. We tested the reliability of a Multi-Layer Perceptron (feed-forward ANN) to automate the process of classification of goat kid calls according to individual identity, group membership and maturation in this species. Vocalisations were obtained from 10 half-sibling (same father but different mothers) kids, belonging to 3 distinct social groups. We recorded 157 contact calls emitted during first week, and 164 additional calls recorded from the same individuals at 5 weeks. For each call, we measured 27 spectral and temporal acoustic parameters using a custom built program in the Praat software. For each classification task we built stratified 10-fold cross-validated neural networks. The input nodes corresponded to the acoustic parameters measured on each signal. ANNs were trained with the error-back-propagation algorithm. The number of hidden units was set to the number of attributes + classes. Each model was trained for 300 epochs (learning rate 0.2; momentum 0.2). To estimate a reliable error of the models, we repeated 10-fold cross-validation iterations 10 times and calculated the average predictive performance. The accuracy was 71.13±1.16% for vocal individuality, 79.59±0.75% for social group and 91.37±0.76% for age of the vocalising animal. Our results demonstrate that ANNs are a powerful tool for studying vocal cues to individuality, group membership and maturation in contact calls. The performances we achieved were higher than those reported for similar classification tasks using classical statistical methods such as Discriminant Function Analysis. Further studies, investigating the reliability of these algorithms for the real-time classification of contact calls and comparing ANNs with other machine learning techniques are important to develop technology to remotely monitor the vocalisations of domestic livestock.

How are dairy cow accelerometer activity changes related to visual gait score changes?

V.M. Thorup[1,2], P.E. Robert[1,2], L. Munksgaard[3], H.W. Erhard[1,2] and N. Friggens[1,2]
[1]AgroParisTech, UMR 791 Modélisation Systémique Appliquée aux Ruminants, 16 rue Claude Bernard, 75005 Paris, France, [2]INRA, UMR 791 Modélisation Systémique Appliquée aux Ruminants, 16 rue Claude Bernard, 75005 Paris, France, [3]Aarhus University, Dept. of Animal Science, Blichers Allé 20, 8830 Tjele, Denmark; vivi.thorup@agroparistech.fr

Lameness is traditionally detected by dairy farmers assessing the gait scores (GS) of cows visually. Yet, gait scoring is subjective and laborious, and most farmers underestimate lameness prevalence. Accelerometers measuring cow activity may help farmers to detect lameness, because lameness changes activity, but the association between GS changes and accelerometer activity changes is unknown. We investigated the association between GS that increased or decreased by 1 and their corresponding accelerometer activity changes. For our preliminary analysis, 145 lactations by 142 loose-housed Holstein cows from four commercial Danish farms were available. Farms were visited 5 to 7 times in 14 months. Four trained technicians visually scored walking cows on a scale from GS1 (normal gait) to GS4 (obviously or severely lame; the few original GS5 observations were grouped with GS4). Activity data were derived from leg-mounted accelerometers (IceTag3D, IceRobotics, Edinburgh, UK). Individual activity means were calculated for each week prior to a GS. Activity changes between two adjacent GS were calculated. According to initial GS, the 118 increasing GS changes were distributed as follows: GS1→2: 48, GS2→3: 34, and GS3→4: 36, and the 86 decreasing GS changes were distributed as follows: GS2→1: 47, GS3→2: 22, and GS4→3: 17. Changes in walking duration (ΔWalk, min/day), standing duration (ΔStand, min/day) and summed acceleration in three dimensions while lying (ΔALie, m/s^2) were analysed (ANOVA) separately for increasing and decreasing GS changes. Initial GS, farm and the interaction were explanatory variables. Results are reported as LSM (SE). ΔWalk: 0.73 (1.00) and 0.15 (0.94) min/day, ΔStand: -17 (16) and -14 (14) min/day, and ΔALie: 21 (26) and 13 (22) m/s^2 for decreasing and increasing GS, respectively. In other words, the magnitude of the activity changes did not differ significantly for decreasing and increasing GS, bearing in mind the large variability in the data. Activity changes were unaffected by farm and initial GS, except when GS decreased, ΔStand differed by 81 (37) min/day ($P<0.05$) between initial GS2 and GS3, and ΔStand differed by -116 (44) min/day ($P<0.01$) between initial GS3 and GS4. In conclusion, our accelerometer activities changed similarly across farms. ΔWalk and ΔALie, but not ΔStand, changed similarly across GS changes both during increasing and decreasing lameness.

An automated system to monitor dairy cows body condition using a time-of-flight camera

J. Salau[1], J.H. Haas[1], A. Weber[1], W. Junge[1], J. Harms[2], U. Bauer[2], O. Suhr[3], S. Bieletzki[3], K. Schönrock[3] and H. Rothfuß[3]
[1]Christian-Albrechts-University Kiel, Institute of Animal Breeding and Husbandry, Olshausenstraße 40, 24098 Kiel, Germany, [2]Bavarian State Research Center for Agriculture, Institute for Agricultural Engineering and Animal Husbandry, Prof.-Dürrwaechter-Platz 2, 85586 Poing-Grub, Germany, [3]GEA Farm Technologies, Siemensstraße 25-27, 59199 Bönen, Germany; jsalau@tierzucht.uni-kiel.de

During early lactation dairy cows experience negative energy balance which leads to more body tissue mobilization. As enduring negative energy balance strongly affects health, fertility, and performance, the body condition should be monitored systematically and accurately. The layer of subcutaneous fat which is bounded by skin and the fascia trunci profunda located at the gluteus medius muscle, respectively the longissimus dorsi muscle is called backfat. Measuring the backfat thickness (BFT) with ultrasound is highly reflective of the body condition. The present study introduces a 3D camera based system for estimating BFT without separation from the herd, fixation, or touching, which are stress factors for the cow. A Time-Of-Flight camera provides 3D data from analysing the phase shift between outgoing and reentering infrared signal. The 3D camera was placed in 2.55 m height above a feeding station and recorded the standing cows' lower back. Software was developed to handle camera setup, calibration, animal identification, image acquisition and sorting, segmentation, determination of the region of interest (ROI), and information extraction automatically. Sorting and ROI determination had error rates of 0.2% and 1.5%, respectively. Profiles of the digital cow surface were taken along the lines between both hips as well as ischeal tuberosities (respectively tail) and hip. In total 13 camera traits were calculated from these profiles and compared to manually gathered BFT. Data was collected from 96 primiparous and multiparous HF cows from July 2011 to May 2012 at the dairy research farm Karkendamm of Kiel University. BFT was measured with a portable ultrasound generator once a week (n=3,163). The herd showed a mean BFT of 9.71 mm (±3.76). All traits significantly depend on the animal and showed very large effect sizes η^2. According to the coefficients of determination, the precision in measuring strongly varied between cows for digitally as well as manually gathered data, but it was comparable between BFT and camera traits. BFT estimation using the cow as random or fixed effect, lactation week and observation month as fixed effects, and a linear regression on the two traits has been promising. The residuals are nearly normally distributed and their variance is quite homogeneous. The correlation among observed BFT values and estimator is 0.96.

Poster 72

The use of the indicators recorded by the computer management program of the cow herd to assess its welfare

R. Antanaitis[1], A. Kučinskas[1] and J. Kučinskienė[2]
[1]*LSMU VA, Department of Non-infectious Diseases, Tilžės Str. 18, 47181, Kaunas, Lithuania,*
[2]*LSMU VA, Department of Food Safety and Quality, Tilžės Str. 18, 47181, Kaunas, Lithuania;*
ramunas.antanaitis@gmail.com

The aim of this study was to assess how the parameters recorded with the computer herd management program reflect the well-being of cows in the studied farm; upon improvement of the well-being of cows, to observe the change in the recorded indicators. Work equipment: computer herd management program 'Afifarm' (Israel), which records the following indicators during each milking: milk yield (kg); electrical milk conductivity (mS/cm); milking time (min); body weight (kg); activity (steps/h); cases of cow diseases will be recorded to facilitate the assessment of its well-being. The test was carried in 2013 (from 14 May till 29 November) in one Lithuanian farm (600 dairy cows). Milk quantity is measured during each milking using the 'Livestock Performance Control' A4 method. Milk samples from all controlled cows are analysed at Milk testing laboratory once per month. Cows are kept in loose-type cowsheds year-round; there is no seasonality of calving; cows are fed and milked twice a day. First calving cows (120) were selected into a separate group. After changing the keeping conditions, the average milk yield in the first calving cow group increased from 25 to 27 kg/d. Moreover, the lactation curve of the first calving cows did not decline, compared to other lactation cows, where the tendency of declining lactation curve was observed. As observed during the analysis of the change in milking time, upon changing the keeping conditions, the milking time of the first calving cows shortened statistically reliably ($P<0.05$), from 7.4 to 5.9 min. The milking time of the older lactation cows did not change in a statistically reliable manner. Electrical milk conductivity has also changed during the experiment. Upon changing the keeping conditions, electrical milk conductivity of the first calving cows reduced on average from 10 to 9.4 mS, whereas in the case of the second lactation cows the increase of this indicator was observed. Change of the keeping conditions did not have any impact on the change in weight. During the entire test the weight differed statistically reliably ($P<0.001$) among the lactation groups. The cows of 3 and older lactations had the biggest weight (on average 640 kg); the average weight of the second lactation cows was 615 kg, while the average weight of the first lactation cows reached 568 kg. Upon changing the keeping conditions, the activity of the first calving cows increased statistically reliably ($P<0.05$) from the average 130 to 150 steps/h. At the end of the experiment the reduction of the activity could be observed.

Use of precision feeders as a tool to measure feeding behaviour

N. Casal[1], M. Tulsà[1], X. Manteca[2], J. Pomar[3], C. Pomar[4], J. Soler[1] and E. Fabrega[1]
[1]IRTA, Veïnat de Sies s/n, 17121 Monells, Spain, [2]UAB, School of Veterinary Science, 08193 Bellaterra, Spain, [3]UdL, C/ Alcalde Rovira Roure 191, 25198 Lleida, Spain, [4]Agriculture and Agri-Food Canada, 2000, College Street, J1M 0C8 Sherbrooke, QC, Canada; nicolau.casal@irta.cat

Precision livestock farming is a recent concept which promotes the use of technologies to improve productive, economical and environmental efficiency and can also be used to better monitor and enhance animal welfare. In this study, two different precision systems for pigs were tested with the aim of determining the optimal number of animals per feeder for one of those systems. Eighty-four entire males were housed in two barns, each one with two pens provided with a computerized feeder: one pen had an IVOG® station (Insentec B.V., Markness, The Netherlands), and the other had an IPF® feeder (Universitat de Lleida, Lleida, Catalunya). Weekly, the size of the pen was modified with two mobile fences, to test 3 different group sizes: 14 (G14), 21 (G21) and 28 (G28) pigs. Food intake, feeder occupation and feeding rate were assessed in each pen, together with other feeding and social behavioural patterns. Feed consumption increased at the expected rate throughout time (P<0.001). However, no differences were found between the different feeding systems or the different group sizes. No significant differences were found in relation to feeder occupation, but G14 tended to occupy during more time the feeder than the other groups (P<0.1). Feeding rate significantly increased together with size of the group, being the fastest for pigs in the G28 group and the slowest for pigs in the G14 size in both feeding systems (e.g. for IPF station 39.89 for G14, 41.32 for G21, 44,48 for G28; P<0.05). Thus, this study indicates than in the present conditions pigs adjusted their feeding rate to compensate the lower availability of the feeder. Other aspects related to social competence which may influence animal welfare, like skin lesions and fighting behaviour, will be further explored before drawing final conclusions on the adequacy of 28 pigs/feeder. Precision feeders showed to be a good tool to evaluate feeding behaviour.

Poster 74

An innovative tool for on-farm data collection and information sharing

F. Dai, E. Dalla Costa, M. Battini, S. Barbieri, M. Minero, S. Mattiello and E. Canali
Università degli Studi di Milano, DIVET, via Celoria 10, 20133 Milano, Italy; francesca.dai@unimi.it

The development of innovative tools based on existing knowledge is a major interest of the EU policies to facilitate research, to enhance registration, access and re-use of data. Aim of the project was to create a digitalized data collection system to improve the efficiency and reliability of data collection on-farm, low-time consuming, reducing mistakes during data transcription and enabling automatic upload of data to a common server. Open Data Kit (ODK) is a free and open-source set of tools, which manages mobile data collection solutions, developed by the University of Washington, Department of Computer Science and Engineering. The application is available for Android devices. ODK was selected for our purposes due to the easiness of building forms without specific knowledge in computer programming, the possibility to create a virtual server to collect and aggregate data, and therefore the possibility to access data from everywhere. On-farm welfare assessment prototype protocols for horses, donkeys and goats – developed in the framework of the AWIN project (FP7-KBBE-2010-4) – were organized in ODK forms to collect animal-based and management- and resource-based indicators. The application was tested in 22 farms (10 donkey, 2 goat and 10 horse farms) to evaluate feasibility and easiness of use. Our experience demonstrated that the app is friendly to use and practical, does not require a long training period, it is flexible and easy to modify and permits to aggregate and analyse data faster and error-free. By the end of September, the application will be further tested in about 100 farms. Future development is planned to create a dedicated app with the same characteristics that immediately provides a preliminary output of welfare status. Data will be available for further analysis on the server. The app would be useful to the development of an accessible data repository on animal welfare and to increase data and knowledge accessibility to all European countries. The authors thank the EU VII Framework program (FP7-KBBE-2010-4) for financing the Animal Welfare Indicators (AWIN) project.

Experimental approach for continuous measurement of ECG, blood pressure and temperature in cattle

I. Van Dixhoorn[1,2], D. Anjema[1], H. Reimert[2], J. Van Der Werf[2] and M. Gerritzen[2]
[1]*Wageningen UR, Central Veterinary Institute, Houtribweg 39, 8221 RA Lelystad, the Netherlands,* [2]*Wageningen UR, Livestock Research, Edelhertweg 15, 8219 PH Lelystad, the Netherlands;* ingrid.vandixhoorn@wur.nl

Improving welfare of cattle during transportation and the slaughter procedures are issues that deserve great attention. Especially the physiologic impact of these processes on animals are under discussion. Therefore, measuring and quantifying the changes of physiological signals during transport and slaughter are of great interest. Obtaining longitudinal characteristics of physiological signals in animals, without interference, is difficult. An individualized longitudinal approach was developed to measure ECG, blood pressure, activity and temperature in cattle under varying conditions such as transport and the slaughtering. Cows were equipped subcutaneously with telemetric transmitters for continuous measurements of physiologic signals without interfering the animals. Nine healthy dairy cows were surgically equipped with implantable telemetry transmitters with transmission range of 3-5 m (Data Science International® DSI®). A data exchange matrix and 12 receivers for large animals were used for data acquisition. During surgery, the animals were placed in a claw trimming box, with a hammock under the abdomen to prevent them from lying down. The cows were sedated with Domidine® and the surgical site was locally infiltrated with Lidocain 2%. The body of the transmitter was placed subcutaneously in the jugular groove. The systemic blood pressure catheter was inserted through a purse string suture in the carotid artery in cranial direction. The positive ECG lead was fixated near the transmitter. The negative ECG lead was inserted through a purse string suture in the jugular vein in caudal direction. The waveform morphology was monitored and optimized during insertion. Tissue glue (3M™ Vetbond™) was used to seal the arterotomy site and prevent hemorrhage. Muscle layers, sub cutis and skin were closed separately. Measurements started directly after surgery for a period of 5 weeks including the period of transport to the slaughter house and slaughter procedure. All cows recovered from surgery within one day. Wounds swelling was seen for only 2 days. No subcutaneous infections were seen post mortally. All transmitters delivered good ECG signals during the complete experimental period. Blood-pressure signals were accurate in 6 out of 9 animals. Accurate signals were seen during transport and the slaughter process. Conclusion The experimental set up provides a basis for accurate acquisition of physiological signals in cattle for long periods of time without interfering and during transport or slaughter processes. The surgery causes only temporarily discomfort to the animals.

Considering normal variation in the gait of non lame cows when developing a lameness detection tool

A. Van Nuffel[1], G. Opsomer[2], S. Van Weyenberg[1], K.C. Mertens[1], J. Vangeyte[1], B. Sonck[3] and W. Saeys[4]
[1]*Institute for Agricultural and Fisheries Research (ILVO), Technology and Food Science Unit, Burg. Van Gansberghelaan 115 bus 1, 9820 Merelbeke, Belgium,* [2]*Ghent University, Department of Reproduction, Obstetrics and Herd Health, Faculty of Veterinary Medicine, Salisburylaan 133, 9820 Merelbeke, Belgium,* [3]*Institute for Agricultural and Fisheries Research (ILVO), Animal Sciences Unit, Scheldeweg 68, 9090 Melle, Belgium,* [4]*Katholieke Universteit Leuven, Division Mechatronics, Biostatistics and Sensors (MeBioS), Kasteelpark Arenberg 30 bus 2456, 3001 Leuven, Belgium; annelies.vannuffel@ilvo.vlaanderen.be*

Automatic lameness detection requires that every lameness alert corresponds to changing gait or behaviour related to lameness. However, due to the large variation in gait variables between and within cows, the risk of false alarms is considerable. Moreover, apart from painful claw lesions or leg injuries, alterations in the gait of cattle can also be caused by diseases such as mastitis or abomasal displacement due to a general feeling of sickness or painful body parts (e.g. painful udder in case of mastitis). On the other hand, the larger part of the variation in gait variables could be attributed to differences between cows. This suggests that cow specific features can influence gait. To function accurately, a lameness detection system should be able to distinguish between normal variation due to non-disease related cow factors, environmental and management factors, and variation due to lameness. Therefore, the effects of environmental (rain, dark environment), and non-disease related cow factors (age, production level, lactation and gestation stage) on the specific gait variables measured by the GAITWISE were investigated in 30 non lame Holstein cows for a period of 5 months. Darkness did not significantly influence any of the specific variables in the final model. In rainy weather, however, cows did take smaller and more asymmetrical strides. In general, older cows had a more asymmetrical gait and they walked slower and with more abduction. Production level only significantly influenced the force distribution between left and right legs. As with age, lactation stage, calculated as days in milk, showed significant association with asymmetrical and slower gait and less step overlap which became negative towards the end of lactation. The latter could possibly be attributed to the size of the udder and fetus that both change with age, lactation and gestation stage. The results in this study demonstrate that – for non-lame, healthy cows – several gait variables measured by the GAITWISE depend on the stage of lactation or stage of gestation. This should be taken into account when analysing individual cow gait data for the detection of alterations caused by lameness.

Development of kinect based self calibrating system for movement analysis in dairy cows

J. Salau, J.H. Haas and W. Junge

Christian-Albrechts-University Kiel, Institute of Animal Breeding and Husbandry, Olshausenstraße 40, 24098 Kiel, Germany; jsalau@tierzucht.uni-kiel.de

Lameness as well as body condition loss are problems of animal welfare, health, fertility, and productivity. Automatically analysing dairy cows' movement and body characteristics without the need to separate individual animals from the herd and examine them, reduces stress for the animals and increases accuracy as well as objectivity. In this study a system based on Microsoft Kinect 3D cameras is going to be developed to assess animal welfare with methods of Computer Vision. As a prototype, a gate with pass line height 2.05 m (total height 2.15 m) and passage width 2.38 m (total width 2.54 m) was constructed and equipped with six Kinects. To observe several steps of cows passing through the gate, two crossed Kinects were mounted in both upper corners having a combined horizontal field of view of 110°. Therefore, they are able to record the approaching as well as the leaving cow. Additionally, from both sides the udder, ankle joints, and lower rump are monitored by Kinects in approximately 0.8 m height. A firmly installed future construction will provide guidance to keep the freely walking cows on the centre line, but provisionally a Holstein Friesian cow was led through the framework by ropes for test recordings at Kiel University's research farm Karkendamm in August 2013. Software to simultaneously operate multiple Kinects has been written in C++ and a data format suitable for analysis has been developed. Synchronization of the recordings from the six cameras has successfully been managed, and algorithms for calibration of the system in terms of determining cameras' heights and inclinations, background scenery, and sorting the images according to 'cow' or 'no cow' have been implemented. In addition, the segmentation of 'cow' images, automatically marking the claws and analysing the animal's moving direction has already been developed. Reasonable next steps are the analysis of trajectories of claws and joints as well as the description of posture for the detection of gait changes, and of course the extraction of body traits to monitor changes in body condition. The aspired system will provide many opportunities to automatically gather welfare measurements. Furthermore, the almost complete 3D information about the animal makes the development of new measurements in animal welfare thinkable.

How do fattening pigs spend their day?

J. Maselyne[1,2], W. Saeys[2], B. De Ketelaere[2], P. Briene[3], S. Millet[4], F. Tuyttens[4] and A. Van Nuffel[1]
[1]ILVO (Institute for Agricultural and Fisheries Research), Technology and Food Science Unit, Burg. van Gansberghelaan 115 bus 1, 9820 Merelbeke, Belgium, [2]KU Leuven, MeBioS, Kasteelpark Arenberg 30 bus 2456, 3001 Heverlee, Belgium, [3]HoGent, Faculty Nature and Technology, Brusselsesteenweg 161, 9090 Melle, Belgium, [4]ILVO, Animal Sciences Unit, Scheldeweg 68, 9090 Melle, Belgium; jarissa.maselyne@ilvo.vlaanderen.be

Behaviour in pigs is closely related to their health and welfare status. Health and welfare problems can increase aggressive behaviour or decrease feeding and drinking behaviour. Lately, much attention is going towards automated measurements of behaviour aimed to identify problems in the barn. These systems need validation based on observations of behaviour as 'gold standard'. We are currently testing the usefulness of a system with receivers mounted around the feeding and drinking place. In the present experiment, we observed the behaviour of 3 healthy barrows and 3 healthy gilts in a group of 59 during 3 days (approximately 12, 18 and 24 weeks of age) between 7:00 and 21:00 h. The 6 pigs were marked individually and observations were continuous. The pigs were housed in an automatically ventilated barn with partially slatted floor, 2 commercial round feeders (dry feed) and 4 drinker nipples. Weight of the pigs was 31.8±4.7 kg, 56.2±9.0 kg and 93.3±12.2 kg (mean ± standard deviation) on the 3 observation days. Average daily gain of the pigs between the first and last observation day was 0.73±0.11 kg/day. The pigs spent 68.14±7.28% of the daytime inactive, 13.40±4.47% exploring, 6.90±1.23% feeding, 4.90±4.00% social, 3.37±1.71% active, 2.24±1.08% aggressive and 1.05±0.45% drinking. Total activity was high around 12:00 and 14:00 h, but activity was even larger during the evenings. During 27.3±12.1% of the time, the pigs were close to one of the feeders (<1 m distance, corresponding to 16% of the pen). About half of the agonistic behaviour and one third of the social, exploratory and active behaviour occurred close to the feeder. In the daily schedule of the pigs differences between pigs and days were visible. Bouts of inactivity were alternated by bouts of activity. Of the 1962 feeding bouts, 42.41% was immediately followed by aggression and 12.23% was immediately followed by exploratory behaviour. Using these observations a sensor for measuring feeding behaviour will be validated.

Can an automated restraint test help analysing personality in cattle?

K.L. Graunke[1,2], J. Langbein[2], G. Nürnberg[3], D. Repsilber[3] and P.-C. Schön[2]
[1]University of Rostock, Faculty of Agricultural and Environmental Sciences (AUF), PHENOMICS office, Justus-von-Liebig-Weg 6, 18059 Rostock, Germany, [2]Leibniz Institute for Farm Animal Biology (FBN), Institute of Behavioural Physiology, Wilhelm-Stahl-Allee 2, 18196 Dummerstorf, Germany, [3]Leibniz Institute for Farm Animal Biology (FBN), Institute of Genetics and Biometry, Wilhelm-Stahl-Allee 2, 18196 Dummerstorf, Germany; graunke@fbn-dummerstorf.de

The importance of personality or temperament in livestock to animal welfare and breeding has gained acknowledgement from people working within these fields. Easy applicable behaviour tests to measure personality are strongly demanded, yet very rare. We therefore aimed to verify whether an easy applicable procedure measuring applied tractive force and related parameters can help analysing personality in cattle (*Bos taurus*). Data of 356 crossbreed calves, tested at 90 and 91 dpn, respectively, in a newly developed automated restraint test (pulling test), was correlated with multidimensional personality types retrieved from a novel-object test (NO) including physiological measures of heart rate variability (HRV). The behaviour parameters from the NO were condensed to two principal components with a principal component analysis (PCA). The calves were divided into nine personality types (SC) according to their PC scores. The HRV-measure RMSSD/SDNN-ratio and SC of the calves in the NO were correlated with a generalised linear mixed model (The MIXED Procedure, SAS 9.3, SAS Institute Inc., USA) to parameters of the pulling test. Weight and sex had no influence on the pulling-test parameters tractive force, holding force, dwindling force, and total force, but on number of pulls (weight: $F=4.27$, $P=0.040$) and maximal tractive force (weight: $F=48.6$, $P<0.001$; sex: $F=4.02$, $P=0.046$) with heavier calves pulling more often and with more maximal tractive force, and with female calves pulling with greater maximal tractive force. Tractive force was significantly influenced by the RMSSD/SDNN-ratio ($F=4.23$, $P=0.041$). SC tended to influence the parameters of the pulling test except for holding force and maximal tractive force (tractive force: $F=1.84$, $P=0.070$; dwindling force: $F=1.81$, $P=0.075$; total force: $F=1.84$, $P=0.070$; number of pulls: $F=1.74$, $P=0.090$). We found the candidate measure tractive force for further investigation in developing an easy applicable automated test for measuring cattle's personality on a large practice scale. Yet, it is questionable, whether a complex structure such as personality can be sufficiently measured by one single test alone.

Poster 80

Towards an automated welfare assessment for broilers

E. Koenders and E. Vranken
Fancom, Research, Wilhelminastraat 17, 5981 XW Panningen, the Netherlands;
ekoenders@fancom.com

There is a growing global awareness of good welfare conditions in animal production, resulting in a need for more precise monitoring. Animal welfare has many different aspects and is therefore difficult to measure. Observations by ethologists is a standard procedure for research purposes, but is very expensive for practical application: after a visit to a farm (approximately 3 h/shed) the ethologist is able to deliver a welfare score over different welfare principles, such as defined in welfare protocol based of the 5 freedoms, such as Welfare Quality®. This is a time consuming and complicated process and results only in a temporary assessment of the welfare. Today, automatic monitoring and controlling techniques, such as precision livestock farming, are becoming more and more important to support the farmer in managing the production process. These new technologies offer possibilities to develop full automatic on-line monitoring of the different welfare aspects. For most of the welfare criteria (absence of hunger and thirst, comfort around resting, thermal comfort, ease of movement and absence of disease), the authors have developed a simplified automated welfare dashboard for broilers based on continuous automated measures of indoor climate (temperature, humidity, light conditions) and the activity and distribution of the birds measured with camera's. In a current EU project, the automated assessment is extended with sound measures and the scores are validated against the expert scores in 5 commercial farms. The first results indicate promising results.

Monitoring welfare in practice on Dutch dairy farms

F.J.C.M. Van Eerdenburg[1], J. Hulsen[2], B. Snel[3] and J.A. Stegeman[1]
[1]Faculty of Veterinary Medicine, Utrecht University, Dept. Farm Animal Health, Yalelaan 7, 3584 CL, the Netherlands, [2]Vetvice, Hoekgraaf 17A, 6617 AX Bergharen, the Netherlands, [3]DLV, Munsterstraat 18a, 7418 EV Deventer, the Netherlands; f.j.c.m.vaneerdenburg@uu.nl

The Welfare Quality welfare assessment protocol° (2009) (WQ) is intensive and takes up to a day to assess a farm. Several other protocols, i.e. Welzijnswijzer (=Welfare Indicator), Koekompas (=Cow Compass) and the Continue welzijns monitor (=Continuous Welfare Monitor), do not require a full day for a single farm and these have been compared with WQ on 60 dairy farms in the Netherlands. Four veterinary practices were asked to make a list of their dairy farmer clients. Each was given a score from good to bad, based on the availability of good quality food & water, quality of housing, health and behaviour. This score was based on the impression of the vets, no assessment was done at this time. Randomly, 60 farms were selected in such a way that in each practice there were 5 good-, 5 average- and 5 bad farms. The farms were visited within 2 weeks for all protocols. The result of the WQ protocol was considered to be the reference and the other three were correlated with WQ. Not only at the level of the end score, but also at principle, criteria and indicator level. The results for WQ were: 3 farms Not Classified, 52 Acceptable and 5 Enhanced, no farm was Excellent. This implies that WQ does not have a proper discriminative capacity. The correlations with the other protocols were very low and not statistically significant. It appeared that mainly the principles Feeding and Behaviour were determining the WQ end score. Therefore, the original WQ was adapted in 3 ways: Drinking water, Integument alterations and QBA. After these adaptations, a new score was calculated and resulted in 22 farms Not Classified, 31 Acceptable and 7 Enhanced, no farms were Excellent. Now all 4 WQ-principles were influencing the end result. The correlations with the other three protocols were still very low at end result level. For some parameters there was a high correlation between the animal based and environmental measures. For example, the number of collisions of the cow with the dividers (during the lying down movement) correlated well with the width of the freestall (r^2=0.63; P=0.03). It was possible to construct a shorter protocol out of the components of the three other protocols tested, with a correlation of 0.88 with the adapted WQ. Measured parameters are: body condition score, water supply, freestall dimensions, softness of the bedding, cleanliness of the cows, access to pasture, cows lying outside the freestall, locomotion score, skin lesions, mastitis, other diseases, and avoidance distance at the feed rack. Execution of this new welfare monitor takes approximately 1.5 h for a farm with 100 dairy cows.

Describing animal welfare: the importance of the index model

L.G. Lawson[1], B. Forkman[2] and H. Houe[2]
[1]Danish Veterinary and Food Administration, Animal Welfare and Veterinary Medicine Division, Stationsparken 31-33, 2600 Glostrup, Denmark, [2]University of Copenhagen, Department of Large Animal Sciences, Grønnegårdsvej 8, 1870 Frederiksberg C., Denmark; lagl@fvst.dk

Animal welfare is multi-dimensional. When presenting the results outside of the scientific community we often need to be able to reduce our numerous welfare measures into one score, creating an animal welfare score. If this score is to be used for comparison over time we will have created an index. There are currently a number of different methods to design such an index available from other sciences, notably economics. The aim of the current paper is to compare the advantages and disadvantages of models based on these scores using a data set organized according to the principles of the Welfare Quality protocol for dairy cattle. Welfare indicator measures and scores are used to calculate elementary (basic) direct, chain and period to period indexes. Tests are conducted to test the degree that the following properties of the index are upheld: (1) Time reversal – if all the data for the two periods are interchanged, then the resulting Score index should equal the reciprocal of the original score index. (2) Transitivity – the chain index between two periods should equal the direct index between the same two periods. (3) Proportionality and identity – if the score of every item is the same as in the reference period, the index should be equal to unity i.e. same as the reference period. (4) Changes in the units of measurement – the index should not be influenced by the changes in measurement. Data from the Danish meat inspection and the welfare control and other relevant data are used to illustrate the consequences of using different models. The consequences of alternative models will be presented resulting in an overview that will enable researchers to better choose an appropriate model among the well-established methods from other sciences. The index model is simple and transparent.

Aggregation of animal welfare on farm level analysed as measurement

K.K. Jensen

University of Copenhagen, Department of Food and Resource Economics, Rolighedsvej, 1958 Frederiksberg C., Denmark; kkj@ifro.ku.dk

This paper presents a framework to analyse aggregation of animal welfare on farm level based on fundamental measurement theory. From this perspective, 'measurement' or aggregation of indicators is then seen as based on an overall comparing relation of pairs of vectors of indicator values (has higher aggregate welfare than). Aggregate welfare on farm level is typically assumed to be a so-called decomposable structure, which means that scores or values on a field of indicators (measured either on individual level or farm level) each are assumed to contribute independently to the aggregate welfare on the farm. Determining what the overall relation looks like is then a matter of determining the weight each indicator value has in determining aggregate welfare. Analysing aggregation in this way makes it possible to clearly undercover the (value and other) assumptions underlying a specific suggestion on how aggregation should be made. A widespread method of determining weights is the use of expert panels; this has e.g. been used in Welfare Quality. The paper analyses the experiences of using expert panels in Welfare Quality. Aggregation in Welfare Quality was made stepwise from measures to criteria to principles up to overall status. The paper will describe the aggregation method used here and contrast it with possible alternatives. Next, it will describe how the experts were used and contrast it with possible alternatives. What sort of questions were experts used to answer and why were they used in this way? What was the experience with the answers in terms of reliability? By comparing these experiences with relevant experiences from other areas, e.g. uncovering people's preferences, the paper concludes in a tentative characteristic of strengths and weaknesses by using expert panels.

Application of an information theoretic approach to evaluate welfare in turkeys during transport

K. Capello[1], G. Di Martino[1], E. Russo[2], P. Mulatti[1], M. Mazzucato[1], S. Zamprogna[1], S. Marangon[1] and L. Bonfanti[1]
[1]IZSVe, viale dell'Università 10, 35020, Italy, [2]Private veterinarian, via Primo Levi 1/B, 35020, Italy; kcapello@izsvenezie.it

The pre-slaughter transport of poultry is a major welfare concern, as several factors can expose animals to social stress, injuries and thermal discomfort. Heat stress has been observed as a major cause of dead-on-arrival (DOA) incidence for broilers, while similar studies on other productive types such as turkeys are scant in literature. In comparison to broilers, turkeys appear being more prone to heat stress, and may have a lower physiological capacity of oxygenation. The incidence of DOA is considered a valuable tool to assess turkey stress during transport, besides representing a cause of economic losses. A wide array of variables may influence the DOA, requiring complex analytical approaches to best define the most important drivers. Similar issues are commonly dealt though statistical methods such as stepwise regressions. Nevertheless, these approaches may impair the correct interpretation of the results, or may not be completely reliable in explaining the observed phenomenon. We applied an alternative approach, widely used in ecology, to analyse data recorded during transport of turkeys to slaughterhouses. In this retrospective study 51,333 slaughtering batches accounting for more than 42 million turkeys were collected for 3 years (2009, 2010, 2011) in two of the biggest slaughtering plants of the same integrated company in Northern Italy. Data for each batch comprised: travel distance and duration, gender and age of animals, cage type, number of animals per cage, number of cages, genetic types and DOA (reported as percentage of dead animals at arrival). Meteorological data (temperature and relative humidity) per each transport were collected from the archives of Environmental Protection Agency. A series of alternative Mixed Effects Models were built to evaluate the DOA distribution in relation with different combination of meteorological variables, travel distance, cage type and their interactions. Genetic type and gender were included as random effects. Model selection was based on Information Theoretic methodologies based on the corrected Akaike Information Criterion. In the case of uncertainty in model selection, a model averaging algorithm was conducted which allowed combination of the parameter estimates from a selected set of models, considering the contribution of each model as proportional to its likelihood weight. This approach provided higher reliability in the interpretation of the results, and particularly in the quantification of the relative importance of the variables in the models associated to DOA variations.

Welfare indicators identification in Portuguese dairy cows farms

C. Krug, T. Nunes and G. Stilwell

Faculty of Veterinary Medicine, Lisbon University, Avenida da Universidade Técnica, 1300-477 Lisboa, Portugal; denoronhakrug@hotmail.com

The objective of this study was to determine the possibility of identifying dairy farms with poor welfare using a national cattle database. The welfare of dairy cattle was assessed using the Welfare Quality protocol on almost 2,000 adult animals from 24 Portuguese dairy farms. After identifying the farms with poor welfare (level 'not classified' according to the Welfare Quality protocol), 15 potential national welfare indicators were calculated based on a national cattle database (Sistema Nacional de Identificação e Registo de Bovinos, SNIRB). The link between the results on the Welfare Quality evaluation and SNIRB was made using the identification code of each farm. To evaluate the probability of a farm having poor welfare, we created a model using the classifier J48 of Waikato Environment for Knowledge Analysis (WEKA). Five farms were classified as having poor welfare ('not classified') and the other 18 were good. Within 15 potential national welfare indicators, only two, proportion of on-farm deaths and female/male births ratio, were significantly different between farms with good welfare ('enhanced' and 'acceptable') and poor welfare ('not classified'). The classifier J48 created a decision tree based on the indicators proportion of on-farm deaths and proportion of calving intervals higher than 430 days to identify farms with higher risk of having poor welfare. With the model created, it was possible to identify correctly 70% of the farms classified as having poor welfare. It is possible to create a model to detect farms with higher chances of having poor welfare. However, a representative number of dairy farms should be evaluated so that a reliable and accurate model can be created for nationwide use. The national cattle database analysis is a good alternative in the dairy welfare evaluation and it could be useful in helping official veterinary services of any country in detecting farms with higher risk of having poor welfare. This approach could be used in production systems of other animal species.

Poster 86

On farm welfare assessment of nursing sows with reference to normal lactating sows
H. Hansted, T. Rousing, P.H. Poulsen, L.J. Pedersen and J.T. Sørensen
Aarhus University, Department of Animal Science, Blichers Alle, 8830, Denmark;
HelleJ.Hansted@agrsci.dk

During the last decade litter size in Danish sow herds has increased and the number of alive born piglets often exceeds the lactating capability of the sow. It is therefore common practice to use 'nursing sows' for fostering the surplus piglets. A 'nursing sow' is a sow extending her lactation beyond the period needed for nursing her own piglets. A 'nursing sow' will therefore have a longer lactation and consequently stay fixated for a longer period in the farrowing crate. The prolonged lactations in farrowing crates may lead to animal welfare problems. Animal welfare problems such as: increased risk of shoulder ulcers, skin lesions (body and udder), leg disorders and low body condition, are therefore expected. However, the knowledge of 'nursing sow' welfare is scarce. Animal welfare in sows is generally affected by management and we expect therefore that the animal welfare problems of 'nursing sows' may vary between herds. To investigate animal welfare of nursing sows we have chosen to assess animal welfare of nursing sows with reference to a similar group of non-nursing sows in the same herd. The advantages of doing so are that we can take into consideration herd differences in animal welfare indicators. However, when making a comparison between nursing sows and non-nursing sows it should be taking in to consideration that the selection of nursing sows is not random but based on criteria's affecting the welfare outcome. In an on-farm study, a group (10-30) nursing sows and a similar number of non-nursing sows in different weeks of lactation is investigated in 60 sow's herds. On each sow a clinical examination is conducted including body condition, skin lesions on udder and body, and shoulder ulcers on the sows, as well as lesions on and body condition of the piglets. Also management routines as for examples criteria for selecting 'nursing sows' as well as resource measures as size of crates and resting surface is measured for each sow. The differences in prevalence of each of these welfare indicators between nursing and non-nursing sows are tested and differences in the animal welfare related to the aspect studied is presented at the conference.

Assessment of pig welfare using both resource- and animal-based measurements

L.L. Lo
Chinese Culture University, Animal Science, 55 Hwa-Kang Road, 11114 Taipei, Taiwan;
loll@faculty.pccu.edu.tw

Substantial livestock production system relies on high productivity and quality animal welfare. Good housing, good feeding, good health and appropriate behaviour are the four principles defined on Welfare Quality® Assessment Protocol for Pigs, in 2009. EU regulation (Directive 2008/120/EC) ban on sow stalls (except the first 35 days of their pregnancy) takes effect on January 1, 2013 after a transitional period of 12 years. An investigation was conducted to evaluate a series of resource-based and animal-based measures for the assessment of pig welfare in Taiwan. A total of 442 pig farms were invited to participate at the research by sending mails and 65 of these were visited in August 2009 and March 2010. About 68% farmers kept their sows in gestation stall for the whole gestation period. The sizes of the stall were ranged from 1.8-2.8×0.6-1.6 m. Within four types practice: resource status, health condition, feeding and management, and stockmanships, 33 different welfare measures for farms were identified. One of the three digits: -1, 0, and 1 indicating poor to excellent were given based on the corresponded measures. One hundred and seven farms received a score 1 for not clipping or grinding piglets teeth, while ninety-three farms were not practicing tail docking. Teeth clipping is the most probable practice to be quit from the farms. On the other hand, castration of male piglets to avoid the boar taint is the common practice in Taiwan. Farmers who have successors are more willing to attend welfare-related course than those not have successors (50 vs 33.3%). With the increasing demanding of consumers for better care of animals. Results of the study provide the information on current status of pig welfare and might be used to develop a more objectively and detailed measures and criteria for assessment of pig welfare in Taiwan.

National cattle movement database use to assess beef cattle mortality in a sustainable farming view

M. Brscic, C. Rumor, G. Cozzi and F. Gottardo
University of Padova, Department of Animal Medicine, Production and Health, Viale dell'università 16, 35020 Legnaro (PD), Italy; marta.brscic@unipd.it

Farm animal mortality is relevant for both, farm economic losses and environmental impact and its reduction could be a key factor for the improvement of farm sustainability. Mortality data in intensive beef cattle farms are rather unexploited by scientific literature but the use of national cattle movement databases could be a valuable opportunity to monitor it and to identify subsequently the farm critical management points. This study aimed at investigating mortality prevalence from 2010 to 2012 in 30 intensive beef cattle farms in the Veneto region (Italy), one of the major beef production areas at national level. The study was carried out by investigating the official national cattle movement database and by collecting per each farm per year data regarding total number of bulls reared, cattle breed and gender, total number of bulls slaughtered and died, age at housing and at slaughter/death, and stage of the fattening cycle in which mortality occurred. Mortality rate was the proportion of dead animals over the total number of animals reared per year. Data were submitted to descriptive statistics. Mortality data were then submitted to analysis of variance with a linear mixed model that considered farm size class, prevalent breed, and stage of fattening as fixed, farm within year as random effect and year as repeated with the Bonferroni adjustment option. Descriptive data at farm level revealed a very high variability of farm size with 803±1,025, 877±1,149, and 832±1,092 (average ± SD) bulls reared per farm in 2010, 2011 and 2012, respectively. Farm size ranged from about 40 to over 5,000 bulls reared per year and was classed as small (≤300 bulls/year), medium (301-800) and large (>800). Mainly male French beef breed bulls were fattened (44.2% Charolais, 15.3% Limousine, 27% meat crossbreds, and 13.5% other breeds) with females ranging from 0% in 15 farms to 100% in 3 farms. The average housing age was 327±40 days. Mortality rates were 1.12±0.93, 1.27±1.08, and 1.40±1.37 for year 2010, 2011 and 2012, respectively. Animal death occurred mainly in the first half of the fattening cycle and on average after 97.2±50.8 days on feed. Mortality rates were not affected by farm size, prevalent breed and stage of fattening ($P>0.05$). However, the high variability among farms suggest that it would be important to investigate potential risk factors, particularly in farms where mortality rates over 2% occur at a late stage of fattening. It could be concluded that exploitation of national cattle movement databases to assess mortality of beef cattle has a potential for highlighting farms in which efficiency and sustainability should be improved.

Mouth lesions in riding horses: associated bridle characteristics and management factors

K. Van Campenhout, E. Roelant and H. Vervaecke

University College KAHO, Agro- and Biotechnology, Ethology and Animal Welfare, Hospitaalstraat 23, 9100 Sint-Niklaas, Belgium; kim.v.campenhout@gmail.com

Given the relatively high number of observed mouth lesions, we performed a welfare study in ridden horses in Flanders to detect possible risk factors. We examined 110 horses in total of which 47% were housed in public, 28% in private riding stables and 25% at private homes. Our goal was to quantify the number of lesions in the mouth and explore if this number of lesions was associated with specific bridle characteristics and management factors. An SPSS linear regression analysis was performed on: stable type, sex of riders, number of riders, duration of use, frequency of use, competition frequency, discipline, bit material, bit type, bit weight, tightness of the bit, presence of cheek protection, accessory reins, noseband type, width, tightness, height of noseband. Our study included four types of nosebands: English or high noseband (30%), German or low noseband (12%), combined nosebands (52%) and Mexican (6%). 62% of the horses showed lesions in the mouth. There were significantly more lesions in horses housed in public stables versus private homes (P=0.004). We observed that English and German nosebands were associated with a higher number of lesions, the effect of the noseband on the lesions was confirmed with an overall P-value of 0.000 comparing the four nosebands. Post-hoc only a significant difference between the English and combined noseband was found (P=0.001). Longer duration of use, higher frequency of use and four or more different riders were factors associated with a higher number of lesions. The width of the noseband had a significant effect, as fewer lesions were observed in horses tacked with wide nosebands. Tightness of the noseband was a significant predictor, with the highest mean score observed for the medium bands which was significantly different from the tightest bands (P=0.034). Replication of these findings or further refinement of the methodology is required prior to drawing conclusions. The lesion-free horse was typically a horse tacked with a wide (mean 2.7 cm) combined noseband, ridden by one rider, riding on average four times per week during 53 minutes per day.

Tail lesions in conventional German pigs at the abattoir: seasonality and relation to herd size

A.L. Vom Brocke[1], D. Madey[1], C. Karnholz[2], M. Gauly[3], L. Schrader[1] and S. Dippel[1]
[1]Friedrich-Loeffler-Institut, Institute of Animal Welfare and Animal Husbandry, Dörnbergstraße 25/27, 29223 Celle, Germany, [2]University of Natural Resources and Life Sciences Vienna, Department of Sustainable Agricultural Systems, Gregor-Mendel-Straße 33, 1180 Vienna, Austria, [3]University of Göttingen, Department of Animal Science, Albrecht-Thaer-Weg 3, 37075 Göttingen, Germany; astrid.vombrocke@fli.bund.de

Tail biting is a major welfare and economical problem in conventional fattening pig production. Public discussion frequently accuses larger herds of having lower animal welfare levels, or in this case, more tail biting. Furthermore, farmers often point out seasonal influences as tail biting triggers. This study presents first results regarding tail lesions in relation to season and herd size based on data collected at a German abattoir. The study included pigs from 32 conventional indoor farms with fully ventilated housing. Mean herd size was 1,558 fattening places (range 700-4,000). In total, 31,128 mostly tail docked pigs which were slaughtered between 07/12 and 07/13 were automatically photographed from two angles after scalding. From these pictures, three observers scored tail lesions on a 4-point scale: 0=no lesion, 1=slight lesion, 2=major lesion, 3=necrosis. Observers were trained and tested before data collection and at regular intervals in order to ensure sufficient agreement (all PABAK >0.73). For analysis, lesion scores were combined as 'lesion' (scores 1, 2, 3) and 'no lesion' (score 0). Seasonality was tested using Chi-Square tests, while herd size was correlated with overall farm prevalence. Overall tail lesion prevalence was 31.9%. Lesion prevalence significantly differed between all seasons (corrected P<0.001) except between summer and winter (corrected P=1.0; animal numbers for spring, summer autumn and winter were 8,625, 7,955, 8,468 and 6,080, respectively). Lesions were most prevalent in autumn (44.5%) and least prevalent in spring (25.8%). The prevalence was not correlated with number of fattening places per farm ($r_{Pearson}$=-0.19, P=0.310, n=32). Our results indicate no relationship between the observed herd size (700 to 4,000) and tail biting prevalence. However, season seems to have an impact on tail biting activities. Pigs in fully ventilated housing systems should therefore be monitored more closely in periods of high temperature changes, and air conditioning should be adjusted accordingly.

Assessment of welfare in groups of commercially transported pigs: the effects of journey time

M.A. Mitchell[1], P.J. Kettlewell[1], M. Farish[1], K. Stoddart[2], H. Van Der Weerd[3] and J. Talling[2]
[1]SRUC, AVS, Roslin Institute Building, Edinburgh EH25 9RG, United Kingdom, [2]FERA, Sand Hutton, York YO41 1LZ, United Kingdom, [3]ADAS, Boxworth, Cambridge CB3 8NN, United Kingdom; malcolm.mitchell@sruc.ac.uk

A variety of approaches may be adopted for the assessment of animal welfare in relation to transportation. Laboratory based transport simulations, well controlled experimental models and simulations and sampling of commercial journeys have all be employed to determine or predict the effects of specific factors such as journey time and thermal environments. It is important to evaluate and validate such approaches and models for welfare measurement in groups of animals under truly commercial conditions. The present study has attempted to examine the effects of journey time (<2 h to >10 h) on the welfare of pigs by measurements of condition, behaviour and post-mortem carcase indices in a commercial UK slaughterhouse. 84 journeys were examined across the range of journey times during different seasons of the year and on a variety of vehicles. Post transport measures including mortality, injury, lesions, casualty slaughter rates, lameness, behaviours in lairage (during a one hour, post-journey observation period) and meat quality parameters including pHi, pHu and colour were employed to assess the possible welfare status of the pigs upon arrival and to estimate the degree of pre-slaughter stress imposed. Data were analysed and the effects of all factors in the models were determined by GLMM (logit link function), Logistic regression models and REML. The objectives were to determine if journey time had any direct effects upon stress and animal welfare and to identify if any specific upper limit of journey time for safe transportation could be identified. There were no statistically significant indications that journey time had a detrimental effect of the pigs transported under commercial conditions in the range of travel times sampled in this study. It was demonstrated, however, that factors other than travel time may have important impacts in determining the responses of pigs to transport stress and therefore the assessment of the welfare of pigs in transit. These factors include the thermal conditions on the vehicle, the nature of the system in which the pigs were produced prior to transport and interruptions to the journey i.e. stationary periods the most important of which may be the 'standing time' at the abattoir. Whilst the effects of all these factors may be exacerbated by excessive journey times the journey time per se may not be the most important issue when attempting to optimise welfare in transit.

Poster 92

Can welfare assessment and risk factor analysis contribute to welfare improvements in veal calves?

H. Leruste[1], M. Brscic[2], G. Cozzi[2], B.J. Lensink[1], E.A.M. Bokkers[3] and C.G. Van Reenen[4]
[1]Groupe ISA, Equipe CASE, 48 boulevard Vauban, 59046 Lille cedex, France, [2]University of Padova, Department of Animal Medicine, Production and Health, Viale dell'Università, 16, 35020 Legnaro (PD), Italy, [3]Wageningen University, Animal Production Systems Group, P.O. Box 338, 6700 AH Wageningen, the Netherlands, [4]Wageningen University and Research Center, Livestock Research, P.O. Box 65, 8200 AB Lelystad, the Netherlands; helene.leruste@isa-lille.fr

In Europe, minimal standards for the protection of veal calves are defined in EU directives. These standards however do not necessarily guarantee a sufficient level of animal welfare as each farm has specific conditions for housing and management that may cause welfare issues. There is a demand for a scientifically-approved on-farm animal welfare assessment tool. In this study, animal-based measures for the assessment of 5 welfare aspects, i.e. human-animal relationship, abnormal oral behaviours, respiratory, gastrointestinal and locomotion problems in veal calves were developed and assessed for their validity, feasibility and repeatability on a large sample of 174 veal farms across Europe. Factors such as season had an effect on the outcome of the welfare assessment. They should be considered carefully during welfare assessments and should be taken into account when evaluating and comparing farms. After this validation phase, factors imposing a potential risk and factors potentially beneficial for welfare were identified on farms. Twenty-two measures belonging to the 5 welfare aspects were selected for a risk factor analysis. For each measure the best model (with all factors significantly affecting the measure and with the highest R^2) was selected using logistic regression. The best models explained between 16 and 48% of the variance. The factors that were significant in the models were related to several characteristics of the farm such as production and housing system (e.g. number of calves per pen, type of floor), batch of calves (e.g. breed, weight at arrival), management of calves and farmer's experience (e.g. use of babyboxes, year of experience), feed and feed distribution system (e.g. amount of milk powder, type of solid feed). Knowing these factors and understanding their effect on the outcome of welfare measures can help in advising farmers in their effort to improve welfare of veal calves. Risk factors for the welfare of calves could be further investigated and risk factor analyses could be implemented once larger data sets become available from future welfare assessments. More research in specific experimental set-ups could be performed, to determine causality between risk factors and welfare measures. This can contribute to the provision of scientifically-sound advice to veal farmers and the improvement of veal calf welfare.

Poster 93

Welfare of horses transported to slaughter in Canada: assessment of welfare and journey risk factors

R.C. Roy, M.S. Cockram and I.R. Dohoo
University of Prince Edward Island, Sir James Dunn Animal Welfare Centre, Department of Health Management, Charlottetown, C1A 4P3, Canada; rroy@upei.ca

There is concern over the welfare of horses on arrival at slaughter plants in Canada after a long journey from the USA. This study was designed to provide quantitative information on welfare issues associated with the transport of 3,940 horses from 150 loads to a slaughter plant in Canada and to understand the association between journey characteristics and welfare outcomes. Five percent of the horses arrived from Ontario (median duration=12 h), but 95% arrived from 5 states in the USA, after median journey durations of 15-36 h. 5% of horses had a body condition score <3 (scale 1-5), <1% were sweating, <1% showed obvious signs of lameness, 97% were alert and 3% were apathetic. The prevalence of injuries was greater in horses from the USA (13%) than in horses transported from within Canada (4%) (two-sample test of proportions, P<0.001), but the prevalence of horses with pre-existing conditions, such as swollen joints and granulation wounds was greater in horses transported from within Canada (6%) than in horses from the USA (1%) (P<0.001). Six horses from the USA (0.16%) arrived in a non-ambulatory condition. In 100 horses from 40 loads studied in detail, 33% had injuries identified by visual assessment (27% superficial, 6% subcutaneous, 5% with swelling and 12% bleeding), 48% had patches of raised skin temperature identified by thermography and 72% had bruising (11% with identifiable bite marks) identified by carcass assessment. A negative binomial model showed that the number of horses per load with an injury increased by journey duration (P<0.005). A general linear mixed model showed that plasma total protein concentration was increased by journey duration (P<0.001) and by summer compared with winter (P<0.001). There was no significant effect of journey duration on blood concentrations of lactate and glucose or on plasma osmolality. As no pre-transport measurements were made, it was not possible to attribute the results definitely to transport conditions. Lairage and slaughter may also have affected some of the results. The welfare assessment and the use of multivariable analytical methods suggested that long journeys were associated with some negative welfare outcomes that could potentially be mitigated by changes to management practices. Journey duration increased the risk of dehydration and the risk of injury e.g. from biting or kicking or from impacts with other horses or the vehicle. Fewer severe welfare problems were identified in this study than in similar studies conducted previously in the USA. In addition, it is possible that changes in traceability procedures subsequent to this study could have improved the welfare of the horses.

Welfare of horses transported to slaughter in Iceland

R.C. Roy[1], M.S. Cockram[1], I.R. Dohoo[1] and S. Ragnarsson[2]
[1]University of Prince Edward Island, Sir James Dunn Animal Welfare Centre, Department of Health Management, Charlottetown, C1A 4P3, Canada, [2]Hólar University College, Department of Equine Studies, Hólar, Hjaltadalur, Iceland; rroy@upei.ca

There is considerable interest in the transportation of horses to slaughter and a need for studies to assess the welfare implications of this practice. Although potential welfare issues have been identified in North America and the EU, there have been no studies on the equine slaughter industry in Iceland. In the autumn, foals and adult Icelandic horses that are not required for breeding are rounded-up and transported for slaughter. In 2010 and 2011, 46 loads of 7-35 horses transported in a non-articulated single deck vehicle for 0.33 to 3.10 h to a slaughter plant in Iceland were studied. On arrival, the horses were kept for 11 to 19 h in lairage pens containing automatic nipple drinkers. Between 1 and 11 horses/load (59 adults and 129 foals) were observed during loading at the farm of origin and these horses were then studied on arrival at the slaughter plant and during lairage, blood was sampled at slaughter and carcases were observed post-mortem. No horses had a pre-existing condition, lameness or body condition score <3 (scale 1-5). No horses were apathetic before transport, but 3% were apathetic after transport. No wounds were observed before transport, but 1.6% of horses had small, superficial bleeding wounds in the hock region after transport. The respiration rate and skin temperature were greater after transport than before transport (Wilcoxon signed rank test, P=0.02). Foals had a significantly higher respiration rate than adults (P=0.02). A linear mixed model showed that blood lactate concentration was significantly increased by age (adult versus foal) and by lairage stocking density. Compared to published normal values, blood lactate concentration was elevated in 100% of adults and foals, 13% of adults and 20% of foals had a blood glucose concentration lower than normal, and 58% of adults and 25% of foals had a plasma total protein concentration greater than normal. Forty-four percent of adult and 17% of foal carcasses had a bruise (1 to 10 cm^2). A logistic regression model identified that adults had a higher risk of bruising than foals (OR 3.7, CI 1.5-9.2). The respiration rate, blood lactate concentration and occurrence of bruising suggested that refinement of the following management practices would potentially be beneficial: improved handling facilities and technique, separation of foals and adults during transportation and reduced stocking density during lairage. It is possible that the mild dehydration observed in adults was associated with restricted access to drinking water during the lairage of lactating mares. However, further studies are required to confirm these suggestions.

The welfare of animals confined in a Free Stall barn located in Brazil

G.M. Dallago[1], M.C.C. Guimarães[2], R.F. Godinho[3] and R.C.R. Carvalho[3]
[1]Universidade Federal dos Vales do Jequitinhonha e Mucuri, Animal Science, Rodovia MGT 367, Km 583, n° 5000 Alto da Jacuba, Diamantina, MG, 39100-000, Brazil, [2]Universidade Federal dos Vales do Jequitinhonha e Mucuri, Diamantina/MG, Agronomy, Rodovia MGT 367, Km 583, n° 5000 Alto da Jacuba, 39100-000, Brazil, [3]Fundação de Ensino Superior de Passos, Agronomy, Av. Juca Stockler, 1130, Bairro Belo Horizonte, Passos, MG, 37900-106, Brazil; gabrieldallago@gmail.com

There is a growing concern throughout the world regarding animal welfare that can be harnessed to become a limiting factor in the marketing of products of animal origin. The assessment of animal welfare is necessary to determine which elements are subject to changes in order to devise strategies aimed at significant improvement the treatment that animals received. As such, we conducted this experiment to assess the welfare of animals kept in a Free Stall barn using the Welfare Quality protocol. The choice of protocol was justified because it was a recent evaluation method developed specifically for the evaluation of intensive husbandry systems. It covers the evaluation of a number of criteria allocated into four different principles that together evaluate aspects related to health, food, facilities where animals are housed, and their behaviour. This allowed us to provide a general assessment that integrated a range of factors related to animal welfare. The experiment was conducted in a Free Stall located in São João Batista do Gloria, Minas Gerais, Brazil. This is a highlighted region because it is an important dairy region of south-eastern Brazil. Multiparous and primiparous animals were housed in the barn. The experimental period occurred between April 22 to 26, 2013. The data collection was performed sequentially and recorded on a form provided by the Welfare Quality protocol. Preliminary results were generated using I-spline mathematical functions which were then treated by Choquet integrals. The final results regarding the level of animal welfare was presented in a four-point scale, ranging from excellent, enhanced, acceptable and not classified. The results for the final evaluation indicated that the animals were housed in an acceptable level of welfare. In addition, it was possible to determine by means of the preliminary results the critical points related to animal welfare that can be improved. They were: expression of other behaviours; absence of pain induced by management procedures; positive emotional state; and expression of social behaviours. It can be concluded that the level of welfare applied to the animals in this study were within good levels. Additionally, we can also conclude that there are issues that deserve future attention to improvement in order to ensure the maximum welfare for these animals.

Is on farm mortality of dairy cows affected by grouping?

K. Sarjokari[1], M. Hovinen[1], L. Seppä-Lassila[1], M. Norring[1], T. Hurme[2], P. Rajala-Schultz[3] and T. Soveri[1]
[1]*University of Helsinki, Production Animal Medicine, Paroninkuja 20, 04920 Saarentaus, Finland,* [2]*MTT Agrifood Research Finland, Plant Production Research, 31600 Jokioinen, Finland,* [3]*The Ohio State University, Veterinary Preventive Medicine, 1920 Coffey Road, A100A Sisson Hall, Columbus, OH 43210, USA; kristiina.sarjokari@helsinki.fi*

High cow mortality at a herd level is an indicator of poor welfare, and a financial loss for a farmer. We studied the associations of on farm death (mortality and euthanasia) with cow and herd level features, especially with grouping of heifers and cows in large Finnish dairy herds. Study herds were selected from Finnish dairy herds with more than 80 cows. From the volunteers, 82 herds (median herd size 116, mean milk yield 9,200 kg) fulfilled the inclusion criteria. Data were gathered during a farm visit and from the database. Survival of 10,838 cows was modelled with semi-parametric shared frailty Cox model. The cows were followed for 305 days, from their first calving in 2011. The outcome of interest was time from calving until death on farm. The overall probability of on farm death was 6%; 195 of the cows died unassisted and 453 were euthanized. Most cows died within the very first days in milk; the cumulative percentages being 10, 25, and 50 until days 2, 8 and 33, respectively (median 34 d). The hazard of on farm death (HD) was associated with breed (Hol/Ay HR=1.3), parity (1st/3rd HR=3.2), calving season (Jan-April/Sept-Dec HR=1.3), veterinary treatments for calving difficulty (yes/no HR=2.6), milk fever (yes/no HR=1.8), dilated abomasum (yes/no HR=1.6), or mastitis (yes/no HR=0.36). DH was lower if the barn had both single and group calving pens (HR=0.6) compared to group pens. DH was higher if the feed rack was neck rail combined with some separated feeding places, compared to neck rail (HR=2.2) or separate places (HR=1.9) only. DH was greater if the floors of the walking alleys were mixed with slatted and solid concrete floors, compared to solid concrete floors (HR=1.8). Narrower than 340 cm wide walking alleys were associated with bigger HD, compared to 340-370 cm wide alleys (HR=1.3). Wider or narrower stalls, compared to exactly 120 cm wide, had lower HD (HR=0.7). Increased the herd size for more than 10% from previous year led to lower HD compared to other groups, and herd size milking parlour type interaction was significant variable in the model. There was a trend (P=0.08) towards a lower HD if the close up cows were kept with other cows compared to a separate own group. Some previous authors have reported associations with cow mortality and for example herd size, milk yield, milking system, grazing, calving season, parity and breed of the cows. More research is needed on risk factors for cow mortality at herd level.

Poster 97

Evaluating welfare quality in Hungarian dairy herds

V. Jurkovich, B. Fóris, P. Kovács, E. Brydl and L. Könyves
Szent István University, Faculty of Veterinary Science, Department of Animal Hygiene, Herd Health and Veterinary Ethology, István utca 2, 1078 Budapest, Hungary; jurkovich.viktor@aotk.szie.hu

The aim of our study was to assess animal welfare quality in Hungarian dairy herds. The Welfare Quality protocol was used in 15 lactating dairy farms, covering about 10 thousand animals, to assess the animal welfare situation in our country. The protocol has four main welfare areas and they are further divided into sub-areas. These sub-areas determine the required tests to get a reliable picture on the welfare status of the animals. The areas for testing are the following: good feeding, good housing, good health and appropriate behaviour. Depending on the total number of lactating cows, from 15 to 20% of the animals were examined individually, according to protocol requirements. Out of the 15 tested farms 6 had 'good' and 9 had 'acceptable' rating. When examining the criteria, however, there were major differences. The four 'poor' rates in good feeding principal was due to the inadequate number and cleanliness of the drinkers. Often it would have been enough to achieve a higher score to clean or position better the drinkers. The good housing principle was rated 'good' in almost every farm but details show that this is largely due to the loose system of housing. The 'excellent' score can be achieved by removing the manure more frequently and the appropriate design of boxes. The good health principal was only 'acceptable' everywhere, this is mainly due to the large percent of lameness and disbudding carried out without anaesthesia or analgesia. To measure the appropriate behaviour the protocol uses the qualitative behaviour assessment. Reliable and reproducible measurements are used to assess the behaviour of the animals. An important part of this study point is the assessment of the animal-human relationship, The rude behaviour of the workers and cruel treatment of the animals can be seen from the scores as well as a patient, animal-loving attitude. It can be stated that using the Welfare Quality protocol is reliable to assess the animal welfare status in the Hungarian large scale dairy herds. There is definitely room for improvement regarding animal welfare on Hungarian dairy farms, especially in the fields of lameness, animal and stock hygiene and human-animal relationship.

Does antimicrobial treatment level explain animal welfare in pig herds?

T. Rousing[1], J.T. Sørensen[1] and N. Dupont[2]
[1]*Aarhus University, Animal Science, Blichers Allé 20, P.O. Box 50, 8830 Tjele, Denmark,* [2]*Copenhagen University, Large Animal Sciences, Grønnegårdsvej 2, 1870 Frederiksberg C, Denmark; tine.rousing@agrsci.dk*

There is a growing public concern on antibiotic usage in livestock production based on antibiotic resistance risks in animals and humans. Farmers and veterinarians are consequently concerned that less or delayed treatment will lead to compromised animal welfare. The present paper scrutinizes the question if and if so how antibiotic usage and animal welfare at herd level is linked? It is hypothesizes that the true inter-relation is not strict forward as it relies on treatment threshold that may vary between anything from two extremes: (1) a high medical treatment level keeping the disease level low and hereby is associated with a good animal welfare; and (2) a high medical treatment level indicating a high level of diseased animals and therefore an impaired animal welfare status. The hypothesis testing is further complicated by the fact that the validity and correctness veterinary medicine ordination and usage is uncertain. A cross sectional animal welfare assessment including a total of 25 direct animal behaviour and clinical health measures were carried out in 40 commercial and conventional Danish slaughter pig herds during the period from April 2010 to June 2011. The protocol was inspired by the Welfare Quality Protocol®. The measurements included for example mortality, lameness, swellings, hernia, tail bites, explorative and agonistic behaviour. In each herd equally divided between 3 age groups: 30-40, 50-70 and 90-100 kg, respectively – a total of in average 150 pigs where randomly appointed for group wise behaviour assessments, as well as where in average 130 pigs randomly appointed for an individual clinical examination. Based on measurement weights derived from an expert panel opinion and the prevalences of the 25 measurements aggregated individual farm specific animal welfare indices (AWI's) were calculated. These were associated to the herds' ordinated antibiotic incidences for a period of one year prior to the date of the individual herds' cross sectional on-farm animal welfare assessment. Results showed that even though both AWI's and ordinated antibiotic use varied substantially, no significant relation between the two was found (R^2=0,03 for herd levels of percentage of slaughtered pigs treated with antibiotics per day (measured in average daily doses, ADD) per slaughter pig delivered versus the herds' 'AWI'-score). The results could not support any of the two above mentioned hypotheses. Thus there seems not to be any simple relation between animal welfare and antibiotic usage in slaughter pig herds.

Contribution of farm, caretakers and animal characteristics in the human-dairy cattle relationship

A. De Boyer Des Roches[1,2], L. Mounier[1,2], X. Boivin[1,2] and I. Veissier[1,2]
[1]*Université de Lyon, VetAgro Sup, UMR1213 Herbivores, 1 avenue Bourgelat, 69280 Marcy l'Etoile, France,* [2]*INRA, UMR1213 Herbivores, Centre de Clermont Ferrand, Theix, 63122 Saint-Genès-Champanelle, France; alice.deboyerdesroches@vetagro-sup.fr*

The human-animal relationship is crucial for animal welfare. The caretaker plays a determining role through his/her behaviour and attitudes towards animals. Nevertheless, other factors may intervene such as the genetic of animals or farm equipments. We aimed at assessing the respective influence of characteristics of the caretaker (number of caretakers on the farm, attitude toward animals…), the animals (breed, age…), and the farm (housing system, handling equipment, work organisation…) on the human-animal relationship. The reactions of cows to an approaching observer (Welfare Quality° protocol) were assessed in 120 French dairy farms. At farm level, we used a linear model to analyse the links between the percentage of cows that can be touched and the potential explanatory factors (caretakers', animals', and farms' characteristics). At individual level, we used a mixed model to examine the variability between and within farms in the flight distance of cows (with farm as random factor). The first model explained 29% of the variability between farms in the percentage of cows touched. Among the 53 factors tested, only 9 were kept ($P<0.10$) in the final model: Factors linked to caretakers accounted for 53% of the variability explained by the model (positive attitude towards gentle contact and negative attitude towards rough contacts, 29%; number of caretakers, 17%; gender, 7%); Calving conditions explained 30% of this variability with cows being more easily approached when they did not calve indoor within the herd and when the caretaker interacted less. The variability in the flight distance (second model) resulted largely from variations between cows within farms (80% of the variability) rather than variations between farms (20%). Our results confirm the importance of positive attitude towards animals – as largely reported in the literature – but suggest that other risk factors, such as conditions of calving, could play also an important role in the development of human-animal relationships. In addition, individual characteristics of cows – that could result from their genetic background or their early life conditions – may play an even more important role.

Using e-learning to improve consistency of professional judgements about EU pig welfare legislation

B. Hothersall[1], L. Whistance[1], H. Zedlacher[2], B. Algers[3], E. Andersson[3], M. Bracke[4], V. Courboulay[5], P. Ferrari[6], C. Leeb[2], S. Mullan[1], J. Nowicki[7], M.C. Meunier-Salaün[8], T. Schwartz[7], L. Stadig[9] and D. Main[1]

[1]*University of Bristol, School of Veterinary Science, Langford, North Somerset, BS40 5DU, United Kingdom, [2]University of Natural Resources and Life Sciences (BOKU), Department of Sustainable Agricultural Systems, Gregor Mendel Strasse 33, 1180 Wien Austria, Austria, [3]Swedish University of Agricultural Sciences (SLU), Department of Animal Environment and Health, Almas Allé 10, 750 07 Uppsala, Sweden, [4]Wageningen University and Research Centre, Wageningen Livestock Research, Edelhertweg 15, 8219 PH Lelystad, the Netherlands, [5]IFIP Institut du Porc, 3/5, rue Lespagnol, 75020 Paris, France, [6]Centro Ricerche Produzioni Animali, Viale Timavo, 43/2, 42121, Reggio Emilia, Italy, [7]University of Agriculture in Krakow, Department of Swine and Small Ruminants Breeding, Al. Mickiewicza 24/28, 30-059 Kraków, Poland, [8]Institut National de la Recherche Agronomique (INRA), UMR1079 SENAH, F-35000 Rennes, France, [9]Institute for Agricultural and Fisheries Research (ILVO), Animal Sciences Unit, Scheldeweg 68, 9090 Melle-Gontrode, Belgium; b.hothersall@bristol.ac.uk*

We produced an online training package in 7 languages, providing a concise synthesis of the scientific data underpinning EU legislation on enrichment and tail docking of finisher pigs. Our aim was to improve the consistency of professional judgements. It defined four essential characteristics: 'edible, chewable, rootable and destructible' for assessing the suitability of different enrichments. 121 participants from over 10 countries (including official inspectors, certification scheme assessors and advisors) completed a quiz twice: Control group participants completed the second iteration before, and Training group participants after, viewing the training. Data were analysed using nested models in MLwiN (Iteration within Person within Country). P values represent significant Iteration (1 vs 2) × Group (Control vs Training) interactions, indicating a divergence between groups following training. Participants rated the importance of modifying the enrichment defined in nine scenarios from 1 (not important) to 10 (very important). Training significantly increased participants' ratings in two scenarios: where wood was provided but not being manipulated (P=0.0004) and where a chain was present and being manipulated (P=0.003). Thus training helped participants identify enrichments that were less likely to achieve compliance. Participants were given nine risk factors for tail biting and rated each from 1 (no risk) to 10 (high risk). Initial mean ratings for 'barren environment' were >9 but nonetheless increased significantly after training (P=0.002). Conversely, training led to moderate decreases in risk ratings for heat stress (P=0.0003) and high stocking density (P=0.005), consistent with information in the training. Scenarios relating to tail docking and management were then described. Training significantly increased the proportion of respondents correctly identifying that a farm with no tail lesions should stop tail docking (McNemar's test; P=0.001). Finally, participants rated the importance of modifying enrichment in three further scenarios. Training increased ratings in scenarios where non-compliance was less obvious: a) tail lesions present; pig provided with but not manipulating (wet, dirty) straw (P=0.01), and b) no tail lesions; chains provided and partly used (P=0.006). Therefore the training increased knowledge, and particularly recognition of enrichments that may be insufficient to achieve compliance.

The impact of training on farm assurance assessor confidence and standardisation in welfare outcomes

M. Crawley[1], I. Rogerson[1], S. Mullan[2] and J. Avizienius[3]
[1]*Soil Association, South Plaza, Bristol, BS1 3NX, United Kingdom,* [2]*University of Bristol, Langford, Somerset, BS40 5DU, United Kingdom,* [3]*RSPCA, Southwater, West Sussex, RH13 9RS, United Kingdom; irogerson@soilassociation.org*

Farm assurance schemes have been developed to assure compliance with animal welfare, food safety and environmental standards, throughout the food chain. The 2010 GB Dairy Cow Welfare Strategy was designed by the industry for the industry and listed the inclusion of welfare outcome measures (WOM) into farm assurance standards as a priority. In October 2013, Red Tractor Assurance Dairy Scheme (which certifies 85% of UK dairy farms), RSPCA Freedom Food and Soil Association Certification incorporated assessing WOM as part of routine farm visits. The WOM were chosen based on their welfare importance, practicality of use within a farm assurance setting and ability to improve the assessment of existing standards. A training package was created which aimed to encourage standardisation between assessors and improve their confidence when assessing and discussing WOM with the farmer. Assessors were trained in identifying WOM along with their etiology and welfare significance using photographs, videos and live animals. At the start and end of the training day, assessors scored lameness, thin or fat cows, cleanliness, hair loss, lesions and swellings on photographs and videos of cows, and completed a questionnaire evaluating their attitudes to and confidence in using WOM and exploring their preferred means of learning. A total of 114 assessors were trained by 6 trainers at 10 locations. When scoring WOM on photographs and videos, the percentage agreement (compared to a gold standard) increased in the post-training assessment for five out of the six measures. 79% of assessors scored a percentage agreement of 70% or higher for all measures post-training compared to 46% in the pre-training assessment. Seventy-seven pre- and 94 post-training evaluative questionnaires were completed. Post training, for every WOM, assessors rated their confidence higher in identifying, understanding the causes of, directing to sources of advice, knowing related standards and raising non-compliances. Confidence in raising welfare as an issue was also rated higher post-training. The training workshop improved assessor consistency in assessing WOM, their confidence in both identifying the measures and their relevance to certification. Questionnaire feedback has been used to improve the design of future training, develop useful aids to assessment and provide a baseline of information. An online web tool has been launched to assist with initial and ongoing training in WOM. In February 2014 all assessors assessed further photographs and videos of cows using this tool to review their standardisation.

Inter-observer agreement for qualitative behaviour assessment in dairy cattle in three different countries

C. Winckler

University of Natural Resources and Life Sciences (BOKU), Vienna, Department of Sustainable Agricultural Systems, Gregor-Mendel-Strasse 33, 1180 Vienna, Austria; christoph.winckler@boku.ac.at

The Welfare Quality® (WQ) protocol aims at assessing farm animal welfare using mainly animal-based measures. One of the measures is qualitative behaviour assessment (QBA), which relies on the ability of human observers to integrate the animals' style of behaving using descriptors, which provide information additional to quantitative measures that is directly relevant for animal welfare. For on-farm use, a fixed rating scale of cattle expression with 20 pre-defined terms was developed. It was the aim of the present study to investigate which degree of inter-observer agreement can be achieved after training in the application of the WQ protocol. During three training courses in France, Finland and Sweden, assessors (n=9, 13 and 10, respectively) were given detailed instructions on the procedure and the use of the rating scale. Additionally, all terms were translated into the local languages and discussed between the attendants. This was followed by training sessions using both video recordings and live observations. Inter-observer agreement was tested using 17 1-min video clips of different herds of dairy cattle. Per country, data were submitted to PCA and scores of the first two dimensions analysed using Kendall Correlation Coefficient W. In all countries, the first principal component (explained variance 31-36%) consistently distinguished between negative and positive mood (,agitated'/,frustrated'/,uneasy' vs ,content'/,calm'/,relaxed'). For the second component (explained variance 17-23%), consistently only terms of ,positive engagement' achieved high loadings (,sociable'/,playful'). Kendalls's W ranged between 0.56 and 0.72 for PC1 and 0.52 and 0.56 for PC2 (all P<0.001, df=16). The results indicate that with few hours of training reasonable to satisfactory agreement can be achieved for QBA in dairy cattle using a fixed rating list, which is independent from the background of the assessors.

Poster 103

Trough or bowl: observers need training for assessing resource as well as clinical parameters

S. Dippel[1], D. Bochicchio[2], P. Haun Poulsen[3], M. Holinger[4], D. Holmes[5], D. Knop[6], A. Prunier[7], G. Rudolph[8], J. Silerova[9] and C. Leeb[8]

[1]Friedrich-Loeffler-Institut, Institute of Animal Welfare and Animal Husbandry, Doernbergstr. 25/27, 29223 Celle, Germany, [2]Agricultural Research Council, CRA-SUI, San Cesario sul Panaro, Modena, 41018, Italy, [3]Aarhus University, Dept. of Animal Science, Blichers Allé 20, 8830 Tjele, Denmark, [4]FiBL, Animal Husbandry Division, Ackerstrasse 21, 5070 Frick, Switzerland, [5]Newcastle University, School of Agriculture, Food & Rural Development, Agriculture Building, Newcastle upon Tyne NE1 7RU, United Kingdom, [6]Beratung Artgerechte Tierhaltung e.V., Walburger Str. 2, 37213 Witzenhausen, Germany, [7]INRA, UMR1348 PEGASE, 35590 Saint-Gilles, France, [8]University of Natural Resources and Applied Life Sciences Vienna (BOKU), Institute of Livestock Sciences, Gregor-Mendel-Str. 33, 1180 Wien, Austria, [9]Institute of Animal Science, Pratelstvi 815, 10400 Prague, Czech Republic; sabine.dippel@fli.bund.de

While the need for training on-farm assessors in clinical animal assessment has been widely recognised, assessment of husbandry resources is still often regarded as self-explanatory. Within the scope of the international project ProPIG, 7 observers from seven countries were trained by an experienced observer (gold standard) to assess 15 clinical and 11 resource parameters in organic pigs in eight countries. The initial plan was to train and test observers before farm visit 1 and then again after one year before farm visit 2. Both trainings were repeated with all observers due to unsatisfactory agreement, resulting in T1a+b and one year later T2a+b. Agreement with the gold standard was calculated as exact agreement for categorical parameters (e.g. drinker type; mean n=11 pens/test and parameter, range 1-34) and Spearman rank correlation for numerical parameters (e.g. number of animals; mean n=9 pens, range 4-28). Median (IQR) pairwise agreements [%] were T1a=83 (40), T1b=90 (29), T2a=92 (43), T2b=100 (11) for clinical parameters, and T1a=100 (25), T1b=100 (40), T2a=100 (23), T2b=90 (33) for resource parameters. Mean Spearman r for clinical parameters were T1a=0.52, T1b=0.76, T2a=0.42 and T2b=0.84 with ranges of -0.69, -0.33, -0.79 and 0.34, respectively, to 1.00. Mean Spearman r for resource parameters were T1a=0.59 (range 0 to 1), T1b=0.71 (-1 to 1), T2a=0.40 (0.30 to 0.49) and T2b=0.25 (-1 to 1). Initial training discussions showed that naïve observers differed in their assessment of resource as well as clinical parameters, and real life assessment together with training materials were needed to successfully train on both sets of parameters. We therefore recommend the inclusion of resource parameters in observer trainings for on-farm assessment in order to assure sufficient observer agreement.

Poster 104

Evaluation of welfare inspections at slaughter in the Czech Republic

E. Voslarova[1], V. Vecerek[1], J. Dousek[2], V. Pistekova[1] and I. Bedanova[1]
[1]University of Veterinary and Pharmaceutical Sciences Brno, Faculty of Veterinary Hygiene and Ecology, Department of Veterinary Public Health and Animal Welfare, Palackeho tr. 1/3, 612 42 Brno, Czech Republic; [2]State Veterinary Administration, Slezska 100/7, 120 56 Prague 2, Czech Republic; voslarovae@vfu.cz

Czech legislation has been harmonised with the relevant European Union legislation before the time of accession of the Czech Republic to the EU (1/5/2004) including rules for official controls. Daily welfare inspections of consignments at slaughterhouses are carried out and consist in the inspection of documentation, compliance with the set conditions of transport (duration of transport, distance, prescribed rest periods), including the inspection of the vehicle, ante-mortem inspection of animals and evaluation of their clinical status, checks of handling the animals at slaughterhouses and the actual slaughtering of animals as well as checks of its effectiveness. Furthermore, comprehensive inspections at slaughterhouses are carried out regularly (inspection of the equipment of slaughterhouses with respect to animal welfare, record keeping, care of animals, lairage, stunning and bleeding procedures, etc.) including evaluation of pathological-anatomical findings at slaughterhouses. The aim of this study was to evaluate the results of comprehensive inspections at slaughterhouses in the Czech Republic from 2005 to 2012. In the monitored period, the number of inspections and inspected animals and the number of inspections in which welfare failures were detected and the number of animals concerned were observed and the relative frequencies (%) were counted for the individual years. The data were obtained in cooperation with the State Veterinary Administration. The results were analysed using statistic software Unistat 5.1. For the evaluation of a long-term development trend, the actual and relative frequencies for the 1st period (2005-2008) and for the 2nd period (2009-2012) were calculated and frequencies of both periods were compared on the basis of a chi-square analysis of contingency tables 2×2. A total of 7,338 comprehensive welfare inspections at slaughterhouses were carried out from 2005 to 2012. Comparing the first and the last four years of the monitored period, the average relative number of inspections in which welfare failures were detected decreased from 6.98% to 3.44% (P=0.0000). Thus, in recent years welfare problems related to slaughter occurred less frequently. Considering the increasing scope of inspections it may be concluded that the welfare of animals at Czech slaughterhouses significantly improved during the monitored period.

The supervision of the pig protection in the Czech Republic in the period from 2007 to 2012

V. Pistekova[1], I. Bedanova[1], E. Voslarova[1], J. Dousek[2] and V. Vecerek[1]
[1]University of Veterinary and Pharmaceutical Sciences Brno, Palackeho 1/3, 612 42 Brno, Czech Republic, [2]State Veterinary Administration, Slezská 100/7, 120 56 Praha 2, Czech Republic; pistekovav@vfu.cz

In the Czech Republic, pig protection and welfare are governed by respective European Union and national legislation. The compliance of individual rules is ensured by relevant supervisory activities which are carried out by competent inspectors of the State Veterinary Administration of the Czech Republic – veterinary surgeons. Their main supervisory activities consist in planned inspections of all pig categories according to type of activity involving these animals – animal husbandry, slaughters, killing, transport by road, transport by rail, transport by air, transport by sea, public performances, trade in animals, experiments, experimental projects, veterinary activities, shelters and rescue stations. The aim of this paper is to evaluate the number of failures in pig protection in the Czech Republic and to assess the trend of their development in the monitored period 2007-2012. In stated time period, the number of inspections with and without deficiencies and inspected animals was observed by the type of pig handling for each year. The results were analysed using statistic software Unistat 5.1. For the evaluation of a long-term development trend, the actual and relative frequencies for the 1st period (2007-2009) and for the 2nd period (2010-2012) were calculated and frequencies of both periods were compared on the basis of a chi-square analysis of contingency tables 2×2. A total of 11,087 inspections were performed between 2007 and 2012 and failures in pig protection and welfare were detected in 542 inspections. 6,090,870 pigs were examined during this number of inspections. The majority of conducted inspections with deficiencies included inspections in animal husbandry, slaughters and transport by roads. The comparison of two monitored periods 2007-2009 and 20010-2012 showed the decrease of supervisions with deficiencies from 5.85 to 3.48% (P<0.01). Our study confirms that both professional and public attention given to the conditions of pig protection and welfare in the Czech Republic brings further partial successes.

Poster 106

Applying the welfare quality protocol in Portuguese dairy cows farms

C. Krug, T. Nunes and G. Stilwell
Faculty of Veterinary Medicine, Lisbon University, Avenida da Universidade Técnica, 1300-477 Lisboa, Portugal; stilwell@fmv.ulisboa.pt

The objective of this study was to assess Portuguese dairy cows' welfare using the Welfare Quality® protocol. The welfare of 24 Portuguese dairy farms, 13 from the centre and 11 from the north of Portugal, was assessed using the Welfare Quality protocol. In total, 1,930 lactating cows were evaluated. Welfare measures were statistically analysed to seek for relationships among them. Within the 24 dairy farms evaluated, one was scored as 'enhanced', 18 were considered 'acceptable' and five were 'not classified' according to the Welfare Quality protocol. The main welfare problems identified were: presence of lesions and swellings mainly in the lower hind limbs and neck/back area; approximately 40% of moderate lameness; no pain management in disbudding calves; non-grazing production systems; insufficient or dirty drinkers; severe dirtiness of the udder and hindquarter; and high percentage of cows lying partially/totally outside the stall. Many measures were correlated to each other, namely: percentage of cows colliding against the cubicles while lying down and percentage of cows presenting swelling in the neck/back; percentage of cows with severe lameness and mortality rate. The welfare of Portuguese dairy farms is generally acceptable, meeting minimal requirements. In order to improve it, cubicles should be altered (number, bedding, dimensions), pain management should be applied when disbudding calves, more attention should be paid to cleanliness (drinkers and pen) and to the number of drinkers per pen.

Success or failure of animal welfare payments in the EU: evidence from Estonia, Scotland and Spain

D. Miller[1], G. Schwarz[2], L. Remmelgas[3], P. Goddard[1], M. Kuris[3], J. Munoz-Rojas[1] and J. Morrice[1]
[1]James Hutton Institute, Craigiebuckler, AB158QH, Aberdeen, United Kingdom, [2]Thuenen-Institute of Farm Economics, Bundesallee 50, 38116 Braunschweig, Germany, [3]Baltic Environmental Forum Estonia, Liimi 1, 10621 Tallinn, Estonia; gerald.schwarz@ti.bund.de

The importance of animal welfare in EU agricultural policies is increasing. More than 500 million Euros were budgeted for animal welfare payments (Rural Development Programme Measure 215) in the EU in the period 2007-2013. Despite the increasing importance little is known about the actual impacts of these payments on animal welfare on EU livestock farms. A lack of suitable indicators and methods restrict the evaluation of the impacts of animal welfare payments. The objectives of this presentation are twofold. First, findings from an evaluation of the impacts of EU animal welfare payments on problem-related and resource-based indicators of animal welfare of farm livestock systems are synthesised and the main challenges for future evaluations derived. Second, the suitability of new methods to improve the robustness of the evaluation results on animal welfare is analysed. The evaluation of animal welfare payments is based on farm surveys of 150 livestock farms in Estonia, Scotland and Castilla y Leon, and a sample of veterinarians in Scotland. The samples of the semi-structured farm surveys have been stratified according to the most important livestock types and production systems in these study areas including cattle, sheep, pig and poultry (egg and broiler systems) and participants and non-participants of RDP measure 215. The veterinarians were selected to validate animal welfare issues on participating farms, changes in numbers of cases on local farms and impacts on vet practice activity. The findings of the survey are discussed with respect to criteria and principles of the EU Welfare Quality project. The assessment of new evaluation methods is based on a review of recent methodological developments and stakeholder interviews with 32 evaluation and animal welfare experts carried out in the FP7 project ENVIEVAL. The results of the farm survey and the statistical analysis of the available monitoring data generally show improvements in problem-related and resource-based indicators on farms receiving animal welfare payments, in particular on cattle and sheep farms in Scotland and Estonia. However, robust causal relationships between the support payment and improvements in the various indicators cannot be confirmed based on the available data. More efforts for monitoring programmes of problem-related indicators on participating and non-participating farms are required to use more advanced econometric methods to address evaluation challenges and to assess net-impacts of animal welfare payments and other support measures.

Content and structure of animal welfare legislation and standards

F. Lundmark¹, B. Wahlberg², H. Röcklinsberg¹ and C. Berg¹
¹Swedish University of Agricultural Sciences, Dept. of Animal Environment & Health, Box 234, 53223 Skara, Sweden, ²Åbo Akademi University, Dept. of Law, Fabriksgatan 2, 20500 Åbo, Finland; frida.lundmark@slu.se

The issue of how to measure animal welfare (AW) has been under discussion for quite some time, and may have an impact on how various requirements are included and structured in different types of AW regulations. For example, the EU Commission has stated that it 'will consider the feasibility and the appropriateness of introducing science-based indicators based on animal welfare outcomes as opposed to welfare inputs as has been used so far'. However, when talking about a general shift from resource- and management- (input) based requirements towards more animal- (output) based requirements, we must first be duly informed about the structure and content of the legislation and standards already in place. In this study we analysed the AW legislation and three private standards in Sweden. The aim was to investigate to what extent these existing regulations are designed as resource-, management- or animal-based requirements, and also to analyse the content of the regulations; in which areas do they have common requirements and in which areas do the requirements differ? We focused on the requirements for on-farm housing and management of dairy cattle, including calves. The method used was text analysis. The results showed that it was mainly the private organic standard that had higher demands than the binding legislation. The legislation was generally more specific than the voluntary standards, but there were also differences within the different regulations in the weighting between various requirements; some paragraphs were stated as being more important than others. The borders between animal-, resource- and management-based requirements were not as distinct as expected; we found numerous examples of requirements that could be classified into more than one of these categories, even within the same sentence. Of the 347 paragraphs classified, 26% were purely resource-based, 29% management-based and 1% animal-based, whereas 29% were a mixture of two categories and 8% a mixture of all three categories. As a result of this mixing, resource-based requirements were found in a total of 55% of the paragraphs, management-based in 59% and animal-based in 24%. The main difference between legislation and private standards was that a larger proportion of the requirements in the standards (69%) were at least partly management-based as compared to legislation (47%). Finally, one must remember the importance of not only content and structure but also the level of compliance with any regulations, for a true influence on AW outcomes. Furthermore, the preventive intention of any AW regulation must be emphasized.

Poster 109

Welfare certification and benchmarking of cattle herds based on mortality data

A. Cleveland Nielsen, H. Kjær Nielsen and T. Birk-Jensen
Danish Veterinary and Food Administration, Animal Welfare and Veterinary Medicine Division, Stationsparken 31-33, 2600 Glostrup, Denmark; acln@fvst.dk

In order to improve the welfare status of cattle herds in Denmark, the DVFA is in 2014 implementing an internet based herd-certification program using within-herd age-group specific mortality data. The program aims at improving the welfare status of cattle by: (1) monitoring mortality as an indicator of welfare; (2) benchmarking herds against the national population based threshold values; (3) using internet-based graphical displays on within-herd mortality for risk communication raising awareness among herd owners and vets on possible herd welfare problems; (4) giving warnings of increased mortality through a colour-box; (5) increasing mandatory vet visits for herds either above the threshold values or having received penalties following a herd welfare inspection by the DVFA. The within-herd mortality for cows and calves is measured as the percentage of dead cows/live cows and a survival analysis of 0-180 day old calves using the KME estimator in a 12 months rolling period. The national threshold values are estimated using population distribution statistics on within-herd population data in Proc univariate SAS®, Enterprise Guide 4.3. Within-herd mortality is considered a good indicator of welfare and the validity of data is considered good due to individual animal registration of cattle in the Danish Central Husbandry Register identifying any cattle within a herd at any time. The within-herd mortalities are shown as both percentage of dead cattle and as percentage of the national threshold value. The latter is also used to give a red, yellow or green colour-box indication on the front page of the internet-based program when logging in, indicating the within-herd mortality in relation to the threshold values. Red being above, yellow above 70% (warning level) and green below 70% of the threshold value. The overall cow mortality in Denmark for 2009 to 2013 was 6.2, 6.1, 5.5, 5.1 and 4.9%, respectively and the calve mortality 7.8, 8.7, 8, 7.8 and 8,1%. The national threshold values for mortality are 14% for cows and 20% for calves and are based on data from 2010 and are awaiting new values in 2014 when the program has been active for some time. The cow mortality has declined in recent years, but the mortality of calves has regrettable not. The national threshold values were implemented by the DVFA in 2010 and the Cattle Industry has set goals for reduction of mortality in cattle. These initiatives might have influenced the decrease in mortality in cows, but left the mortality of calves unaffected. The DVFA program intends to alter this situation reducing mortality in both age-groups, thereby increasing welfare in cattle.

Animal welfare in organic versus conventional sheep and goat farming in Norway

I. Hansen

Bioforsk Nord Tjøtta, Parkveien, 8860 Tjøtta, Norway; inger.hansen@bioforsk.no

Regulations concerning animal health and welfare are stricter in organic than in conventional sheep and goat production in Norway. The organic regulations state that sheep and goats should have: (1) A minimum indoor area of 1.5 m² and an outdoor area of 2.5 m²/animal. There are no such demands in conventional farming, but sheep and goats are commonly housed at 0.7-0.9 m²/animal. (2) In organic goat farming at least half of the indoor surface should be solid; and 3) Organically farmed lambs and goat kids should be fed maternal milk in preference to natural milk (milk without non-approved additives) for a minimum of 45 days. In the conventional regulations 'adequate' amounts of colostrum and milk are demanded. Do these essential regulation differences imply that the welfare of animals in organic farming is better? The purpose of this work was to document differences between animal welfare in organic versus conventional sheep and goat production in Norway. The evaluation is limited to behaviour and production parameters as indicators of welfare, and does not include the health aspect. A literature search based on peer-reviewed articles the past 10 years was conducted. It was documented that: (1) increased indoor space allowance resulted in increased lying time, more synchronized lying behaviour, fewer displacements and higher milk yield in sheep and increased lying time and lower frequency of agonistic behaviour in goats, but no changes in weight gain and milk yield; (2) sheep and goats spent 45-50% of their time outdoors when given access to an outdoor yard during winter, although great individual variations were documented. In goats, play behaviour was only displayed outdoors; (3) goats preferred solid lying floors of rubber or wood in cold climate (-12 °C), but did prefer expanded metal floors in moderate temperatures (+10 °C). A two-level lying area reduced the aggression because low-status goats chose the least attractive spots; and (4) in Norway, all lambs usually suckle their dams for at least four months throughout the grazing period, thus there are no differences between the two farming systems regarding the milk feeding practice. The significance of different milk feeding strategies in goat kids is poorly documented, but studies on lambs and calves show that suckling increases the growth rate, gives better social competence and more exploratory behaviour. The same may be assumed for goat kids. This literature review documents that the regulations in organic small ruminant production in Norway have positive implications for animal welfare, especially regarding behaviour and production indicators. In practice, however, there are few differences between the two farming systems because the animals spend 6 months per year on pasture.

Comparison of selected European animal welfare labels: standards and economic importance

H. Heise, W. Pirsich and L. Theuvsen

Georg-August-University of Goettingen, Department of Agricultural Economics and Rural Develpment, Platz der Goettinger Sieben 5, 37073 Goettingen, Germany; hheise@gwdg.de

Shortcomings concerning animal welfare in intensive livestock production systems have gained increased attention in the media and society as well as political relevance. European consumers react increasingly adverse towards intensive livestock farming. Especially the poultry and pork sectors attract heavy criticism in the media and are, therefore, at risk to lose consumer acceptance. Various research studies have shown a considerable sales potential for products that meet higher animal welfare standards since about 20 to 30% of Western European consumers fully agree that the current standards of meat production are too low. Animal welfare labels are promising options to win this group of consumers for the market segment of animal welfare-friendly meat products. The criteria of different animal welfare labels have sporadically been subject to scientific analyses, but there is still no comprehensive survey of the most important animal welfare labels in Europe. The aim of the present contribution is, therefore, to develop a general set of criteria as a basis for the comparison of the standards underlying selected animal welfare labels. Subsequently, we give an overview of producer prices, market prices, quantities produced as well as market shares of the concepts under review. This is done on the example of animal welfare labels from five European countries. The methodological background of this research project is an extensive literature analysis, the development of a general set of criteria for comparing the chosen labels as well as an in-depth analysis of publicly available data on the selected European animal welfare labels. Furthermore, qualitative interviews with members of the selected labels were done to get additional information. At the current stage of analysis, it can be stated that the labels differ slightly with regard to many criteria. However, they exhibit substantial differences concerning prices and market shares. Preliminary findings show that the labels under analysis strongly focus on animal welfare issues and, thus, respond to the growing concerns of consumers towards intensive livestock farming. The examined animal welfare labels provide producers and producers of certified meat the opportunity to differentiate their products and gain competitive advantages in the market. Labels also provide an opportunity to increase transparency for consumers and, thus, improve the unsatisfying reputation of the meat industry. Furthermore, animal welfare labels can serve critical consumers as quality signals and provide guidance when purchasing food products.

Poster 112

Fish welfare: ethology – new database

B.H.P.S. Studer
fair–fish international association, Zentralstrasse 156, 8003 Zurich, Switzerland;
international@fair-fish.net

How can we judge whether a fish farming method is species-appropriate? We first need to know how a species behaves in its natural habitat and the needs it has which are satisfied there. And this is precisely the problem: ethological studies of fish in the wild are still few and far between... Ethological observation in captivity, especially in the form of comparisons of different farming systems can indeed provide pointers for the improvement of the keeping of fish. However in order for studies on captive fish not to lead to circular reasoning, they need to be «calibrated»: by studies in the habitat from which the species in question originates. At the present time, there is a lack of observations of this kind on fish species kept in aquaculture – a vital research need when considering the welfare of fish raised in aquaculture. Our project: In a long-term project, fair-fish international is aiming to build up a database to complement the leading FishBase.org database. All ethological knowledge to be found on fish in the wild and in captivity should be available here – especially on the 450 species that are already being raised in aquaculture today and about whose species-specific needs the fish farmers still know very little. Our goals: The ethological fish database (1) highlights the research gaps, (2) encourages research and (3) enables ethologically-based answers to questions from practitioners and policymakers. Dr. Rainer Froese (Geomar, Kiel), one of the fathers of www.fishbase.org, the world's leading fish database which covers almost all species, appreciates the idea and warmly welcomes it as a complement to FishBase. He considers our first steps to be an 'excellent start'. We are happy to work closely together with Froese and his team. Description of the project and first examples: http://www.fair-fish.ch/was-wer-wo/wo/international/ethology.html.

Poster 113

Is it possible to assess the compatibility of horses in groups by short-term behaviour observations?

J.-B. Burla[1], A. Ostertag[1], H. Schulze Westerath[1], I. Bachmann[2] and E. Hillmann[1]
[1]Animal Behaviour, Health and Welfare Unit, ETH Zürich, Universitätstr. 2, 8092 Zürich, Switzerland, [2]Agroscope – Swiss National Stud Farm, Les Longs Prés, CP 191, 1580 Avenches, Switzerland; jburla@usys.ethz.ch

The Swiss animal welfare legislation for group housing of horses states that 'in groups of five or more well compatible horses, the minimum required area can be reduced by a maximum of 20%'. The term 'well compatible' is defined as 'no frequent occurrence of aggressive confrontations resulting in injuries'. In practice, the assessment of the compatibility is performed by animal welfare inspectors of the cantonal veterinary offices; however, it is not further specified. Therefore, the question arises how reliable the compatibility can actually be assessed during short-term farm inspections. For this purpose, the social behaviour of 10 groups (5-11 horses) was evaluated by direct observations over 4-6 days. In addition, each group was filmed during a feeding and a competitive feed test. Each video was then randomly sent to 7 experts with years of practical or scientific experience with horses and/or ethology. In total, 28 experts each assessed 5 videos of either feeding or feed test and scored the compatibility of the groups at their own discretion on a continuous scale of 0-10; with 0 being 'not compatible' and 10 'very compatible'. The proportions of socio-positive, threatening and aggressive behaviour were calculated and then compared with the mean expert scoring per video, using a Spearman rank correlation. The mean expert scoring of both the feeding ($\rho=-0.61$, P=0.07) and feed test ($\rho=-0.70$, P=0.03) correlated with the proportion of aggressive behaviour but not with the proportions of threatening (feeding: $\rho=0.04$, P=0.92; feed test: $\rho=0.41$, P=0.25) and socio-positive behaviour (feeding: $\rho=0.006$, P=1.0; feed test: $\rho=-0.16$, P=0.66). When comparing the assessments of the situations feeding and feed test, the mean scoring correlated only weakly ($\rho=0.48$, P=0.17). Further, the individual expert scoring within the same group differed substantially; the inter-observer agreement showed ranges from 1.4 to 8.5. 'Compatibility' as defined in the legislation was most closely reflected by aggressive behaviour. Since these interactions are most likely to cause injuries, focusing on them appears as primarily useful. However, the inter-observer reliability was rather poor both within the same group and between the groups in the different situations, which demonstrated the difficulty of an objective assessment of 'compatibility'. It is therefore questionable whether an objective assessment of the 'compatibility' as part of routine farm inspections is possible at all, and whether a reduction of the minimum area by 20% is justified due to such short-term assessment.

Addressing farm animal welfare in Brazil

M.A. Von Keyserlingk[1] and M.J. Hötzel[2]
[1]University of British Columbia, Animal Welfare Program, Faculty of Land and Food Systems, BC V6T 1Z4, Vancouver, Canada, [2]Universidade Federal de Santa Catarina, Laboratório de Etologia Aplicada e Bem-Estar Animal, Dept Zootecnia e Des. Rural, Rod. Admar Gonzaga, 1346, Itacorubi, 88062-000 Florianópolis, Brazil; mjhotzel@cca.ufsc.br

Over the last decade many emerging countries have established themselves as major players in global food animal production. Among them Brazil is now one of the fastest growing producers and exporters of farm animal products, ranking among the top four world exporters for beef, poultry and pork meats. Increases in production have been achieved largely by adopting intensive confinement systems developed in the industrialized world. Considering the criticisms associated with many of these intensive systems, particularly those associated with restriction of movement, food animal production systems in Brazil may receive similar scrutiny in the future. One challenge facing Brazil is that when animal welfare reform is demanded it may not be afforded the half-century timeline provided in Europe to improve farm animal welfare, increasing the economic risk to farmers. Given the changes that have taken place in the E.U. and other countries, there are lessons that could be learned and applied by Brazil that would no doubt ease or prevent the challenges observed in other countries. To aid the needed transformation, we encourage a multifaceted approach; firstly, in the short term, animal industry groups must quickly work to implement proven animal welfare solutions. In the medium term, governmental organizations need to invest in much needed research, both in the relevant natural and social sciences; this research, which must be tailored to local environmental and cultural conditions, will be needed to guide policy reform and to provide practical solutions to welfare issues, especially those that affect many animals and cause considerable pain and suffering, facilitate natural behaviour and maintain animal health. Moreover, it seems obvious that the continued adoption of management practices that have failed to resonate with societal values elsewhere are at great risk of being unsustainable in Brazil; however, to date there is little scientific information on the attitudes of Brazilians towards issues of farm animal welfare. Thirdly, active engagement of all stakeholders, including agribusiness and associated professionals, farmers, and citizens not involved in animal production when discussing animal welfare standards will be paramount; in its absence, decisions may be made in other arenas, without any input from science and the people whose livelihood depends on agriculture. Throughout this process the training of highly qualified individuals will be key to facilitate the transformational shift needed to address farm animal welfare.

Implementation of the WelFur protocols for farmed fur animals

L. Ahola[1], H. Huuki[1,2], J. Mononen[1,2] and T. Koistinen[1]
[1]University of Eastern Finland, P.O. Box 1627, 70211 Kuopio, Finland, [2]MTT Agrifood Research Finland, Halolantie 31 A, 71750 Maaninka, Finland; leena.ahola@uef.fi

The aim of the European WelFur project was to develop on-farm welfare assessment protocols for farmed foxes and mink. The implementation of these protocols is now being tested in the Finnish Fur Farm 2020 project, on approx. 10% of Finnish fur farms. Here we discuss some of the issues that we have faced during this testing. Since farmed fur animals are most often raised from birth to pelting on the same farm, the welfare assessment requires visiting each farm in all production periods, i.e. in winter (breeding animals), summer (breeding animals with cubs) and autumn (growing juveniles and breeding animals). Within each of these periods, the assessments are done in 1.5-2 months. The periodicity of the assessments affects the implementation of the protocols. First, most employees generally prefer full-time jobs over the part-time job that the work as an assessor, with only 5-6 working months a year, can offer. A solution for this could be to train assessors to assess not only fur animals but also cattle, pigs and poultry, to give them assessment work throughout the year. Furthermore, due to the rapid development of the animals, the assessment results may be affected by the assessment date within each period (e.g. the body condition score of the lactating females in the summer period). This means that, to ensure the farmers' equality, the time windows for the assessments should be even narrower than 1.5-2 months. The winter period with cold weathers and a limited number of light hours is the hardest period from assessors' point of view. The summer period, in turn, coincides with the holiday period, which may make recruiting assessors difficult. Thus, any company planning to offer the assessment services faces challenges in terms of recruitment, the number of assessors needed, and the duration of assessors' employment. Training of assessors by persons involved with the development of the protocols seems to be easy to handle. However, after the training itself, there is a continuous need of support for assessors, mainly due to the possible observer drift, and the fact that very detailed instructions for all different situations found on farms cannot be written in the protocols due to differences in the management routines between the farms. One cannot too much emphasize the importance of good informing of the benefits of the assessments for farmers. The farmers, who have read about the assessments, e.g. from professional magazines, are more willing to accept a visit to their farm. Finally, one must always be prepared to disease outbreaks, like the distemper finding on one Finnish farm in winter 2013, interrupting the winter period assessments in Finland for several weeks.

Poster 116

Improving udder health and cleanliness in Austrian dairy herds by animal health and welfare planning

L. Tremetsberger and C. Winckler
University of Natural Resources and Life Sciences, Department of Sustainable Agricultural Systems, Gregor-Mendel-Straße 33, 1180 Vienna, Austria; lukas.tremetsberger@boku.ac.at

Animal health and welfare planning aims at contributing to improvements in the herd through interventions in a structured way. The aim of the present study was to assess health and welfare of dairy cows and implement farm-specific changes in housing and management in order to improve health and welfare. For this purpose, we visited 34 Austrian dairy farms during the winter housing period. Farm selection was based on the motivation of the farmers to participate in the study. All farms had loose-housing barns and kept on average 34 dairy cows (min-max 24-56), predominantly Austrian Fleckvieh. Milk yield per lactation ranged from 5,879 to 9,883 kg with an average yield of 8,339 kg. Health and welfare state of the dairy cows was assessed according to the Welfare Quality® assessment protocol for dairy cattle. Based on the initial welfare assessment, farm-specific health and welfare plans with measures regarding housing and management were developed. In total, 194 measures were discussed on the farms, mainly in the areas 'udder health', 'metabolic health', 'lameness' and 'cleanliness'. After one year, the welfare assessment was repeated and the implementation of the health and welfare plan was evaluated. The degree of implementation of the measures was 57% across all areas, ranging from 42% to 78% for the areas 'lameness' and 'cleanliness', respectively. For data analysis farms were divided into implementation groups, i.e. farms which had implemented measures targeted at the different areas, and respective control groups. Repeated measures ANOVA revealed significant interaction effects between assessment time and group. While the 'average somatic cell score' slightly decreased in the implementation group, the values for control farms increased (time × group $P<0.01$). Similar changes were observed for 'percentage of cows with somatic cell count exceeding 100,000 cells' ($P<0.01$). Concerning 'cleanliness', the 'percentage of animals with dirty teats' decreased in both groups with a more pronounced reduction in the intervention herds ($P=0.029$). Regarding 'metabolic health', the 'percentage of very fat animals' ($P=0.085$) and the 'percentage of animals at risk for acidosis' ($P=0.061$) tended to decrease in both implementation groups and to remain stable or increase in the control groups, respectively. No significant effects were found for the areas 'lameness', 'social behaviour' and 'integument alterations', which were addressed by a small number of farms with low implementation rates only. These findings support the idea that animal health and welfare planning can be seen as a suitable approach for improving udder health and cleanliness in dairy cattle.

Broiler chicken welfare assessment in certified and non-certified farms in Brazil

A.P.O. Souza, E.C.O. Sans, B.R. Müller and C.F.M. Molento
Federal University of Paraná, Rua dos Funcionários, 1540, 80035-040 Curitiba, Paraná, Brazil;
qualitebr@gmail.com

Brazil is the third producer and the leading broiler chicken meat exporter in the world. Independent certifications are required by some importers in European Union (EU) to guarantee compliance with minimal welfare requisites. Our objective was to compare broiler chicken welfare in certified (C) and non-certified (N) intensive farms in the State of Paraná, Brazil, using the Welfare Quality® protocol. Ten farms in each group were evaluated. They employed natural light, wood shaving litter, automatic feeders, nipple drinkers, sprinklers, exhausters and evaporative cooling system. Birds were male and female Cobb500®, assessed at 42.0 ± 1.5 days of age. Data were transformed in scores that ranged from 0 to 100, where 100 is the best condition. Results were compared by Mann Whitney test at 0.05%, and medians are presented in the order C followed by N. The criterium absence of prolonged thirst differed between groups ($P<0.05$), with scores of 80 (68-84) and 73 (47-85), and consequently good feeding principle differed ($P<0.05$), with scores of 81 (71-86) and 75 (52-86). For the good housing principle, the only measure that differed between groups was litter quality, with scores of 67 (14-67) and 34 (14-67). The stocking density was equal in both groups (32.0 ± 2.8 kg/m^2); this suggests that certification requirements do not differ from density employed in Brazil. Injuries that commonly appear on intensively raised broilers did not differ between groups and presented low scores: overall lameness score was 24 (22-26), 3.6% of birds with gait scores 4 and 5, and foot pad dermatitis score was 32 (18-53), 57.8% of birds showing some level of dermatitis. The results are in agreement with the literature and indicate that these injuries remain critical points for chicken broiler on certified farms. The percentage of abscess, on absence of disease criteria, differed between groups ($P<0.05$), but remained lower than in other studies, with values of 0.00 (0.00-0.03)% and 0.02 (0.00-0.03)%. The qualitative behaviour assessment scores were different ($P<0.05$), with low scores of 29 (28-29) and 28 (27-29), which may be associated to barren environment and low activity due to artificial selection. The certification promoted improvements on access to water, litter quality and QBA, but not on broiler chicken critical welfare issues such as lameness, panting, contact dermatitis and stocking density. For these issues, results suggest that farms had minimum welfare standard regardless of certification, complying with EU law, and that information about animal welfare may be reaching the Brazilian broiler chain mainly by commercial channels. In order to further improve broiler welfare, more rigorous standards should be developed.

Evaluation of farm animal welfare in the Czech Republic

V. Vecerek[1], E. Voslarova[1], J. Dousek[2], V. Pistekova[1] and I. Bedanova[1]
[1]University of Veterinary and Pharmaceutical Sciences Brno, Faculty of Veterinary Hygiene and Ecology, Department of Veterinary Public Health and Animal Welfare, Palackeho tr. 1/3, 612 42 Brno, Czech Republic; [2]State Veterinary Administration, Slezska 7/100, 120 56 Prague 2, Czech Republic; vecerekv@vfu.cz

The aim of this study was to evaluate the results of supervision over farm animal welfare in the Czech Republic from 2005 to 2012. In the monitored period, the number of inspections and inspected animals and the number of inspections in which deficiencies were detected and the number of animals concerned were observed and the relative frequencies (%) were counted for the individual years. The data were obtained in cooperation with the State Veterinary Administration. The results were analysed using statistic software Unistat 5.1. For the evaluation of a long-term development trend, the actual and relative frequencies for the 1st period (2005-2008) and for the 2nd period (2009-2012) were calculated and frequencies of both periods were compared on the basis of a chi-square analysis of contingency tables 2×2. A total of 75,038 inspections were carried out in the framework of supervision of farm animals from 2005 to 2012. Comparing the first and the last four years of the monitored period, the annual average number of inspections decreased from 11,061 to 7,699, whereas the average relative number of inspections in which deficiencies were detected increased from 4.36 to 5.18% (P=0.0000). It resulted from the fact that in recent years, the core supervisory activities consisted in planned inspections. The animal welfare inspections carried out at farm animal establishments were planned based on the centrally conducted risk analysis updated always for the given year. The rate of risk was assessed based on the criteria such as the number of animals in the establishment, the species of farm animals kept, the date of the last inspection, the deficiencies detected during the inspections, and the number of animals moved to slaughterhouses and rendering plants in the previous year. The Plan of Inspections was continuously monitored and modified. The number of planned inspections was intentionally reduced so that there was more space for unplanned inspections performed in justified cases. Stress was put on the quality of conducted inspections and the conduct of follow-up inspections in the event deficiencies were detected. Thus, over and above the planned inspections also unplanned and extraordinary inspections took place, in response to changes made and new requirements imposed in the course of the year or as follow-up checks of the fulfilment of instructions imposed in order to remedy the identified deficiencies. The long-term monitoring proved efficiency of transition from global inspections to specifically targeted inspections based on the risk analysis.

A pilot project of an on-farm dairy welfare assessment program (2012-2013)

L. Pierce[1], E. Vasseur[2], P. Lawlis[3], C. Ramsay[1] and J. Wepruk[1]
[1]National Farm Animal Care Council (NFACC), Box 5061, Lacombe, AB T4L 1W7, Canada,
[2]University of Guelph, Department of Animal and Poultry Science, 31 rue St-Paul C.P. 580, Alfred,
ON K0B 1A0, Canada, [3]Ontario Ministry of Agriculture and Rural Affairs, Unit 1-401 Lakeview
Drive, Woodstock, ON N4T 1W2, Canada; downtoearth27@gmail.com

As assurances of good animal welfare become more important to consumers and retailers, it becomes important that robust animal welfare assessment programs are developed that accurately reflect the welfare status of animals. In Canada, Codes of Practice are nationally developed guidelines for the care and handling of farm animals, and are developed according to a process established by the National Farm Animal Care Council (NFACC). As a natural extension of this Code development process, NFACC facilitated the development of the Animal Care Assessment Framework – a nationally coordinated process for developing animal care assessment programs. Dairy Farmers of Canada volunteered to develop an on-farm animal care assessment program using the protocols outlined in a draft framework. A broad range of stakeholders were brought together to develop the assessment program based on the Code of Practice for the Care and Handling of Dairy Cattle (2009). Code Requirements are assessed using a combination of animal-, resource- and management-based measures. Animal-based measures were chosen with high inter-observer reliability based on research carried out on Canadian dairy farms. The draft assessment program was test piloted on farms across Canada by several types of assessors (i.e. certified auditors, veterinarians, dairy industry support staff). Results of the pilot showed that participating producers understood the need for a national animal welfare program in the dairy industry and that the goals of the program reflected producer expectations with respect to an animal welfare assessment program. Producers supported the inclusion of animal-based measures in an on-farm assessment and a majority of producers who participated in the test pilot agreed that body condition and lameness should be assessed. The piloting of the draft dairy welfare assessment provided important feedback to those developing the program and contributed to the successful completion of the Animal Care Assessment Framework and the development of an on-farm animal care program for the Canadian dairy industry.

An advisory tool to help producers improve cow comfort on their farm

E. Vasseur[1], J. Gibbons[2], J. Rushen[3], D. Pellerin[4], E. Pajor[5], D. Lefebvre[6] and A.M. De Passillé[3]
[1]University of Guelph, Alfred, ON, Canada, [2]DairyCo, England, England, United Kingdom, [3]University of British Columbia, Agassiz, BC, Canada, [4]Université Laval, Québec, QC, Canada, [5]University of Calgary, Calgary, AB, Canada, [6]Valacta Inc., Sainte-Anne-de-Bellevue, QC, Canada; vasseur.elsa@gmail.com

Effective management and an appropriate environment are essential for the health and well-being of dairy cattle. Codes of Practice provide guidance to producers for the welfare of cattle in their care. New Canadian recommendations have been established. The objectives of this study were to develop an on-farm evaluation tool to help dairy producers evaluate how well they are doing and to identify management and environment modifications to improve dairy cow comfort. The evaluation tool addressed critical areas of dairy cow comfort, including accommodation and housing (stall design, space allowance, stall management, pen management, milking parlour and transfer alleys), feed and water (body condition scoring, nutrition), and health and welfare (lameness, claw health and hoof-trimming). Targets of good practices were identified from the requirements and recommendations of Code of Practice. Each farm received a score for each target, ranging from 0 (target not reached) to 100 (target reached). One hundred tie-stall and 110 free-stall farms were surveyed in three provinces of Canada (Quebec, Ontario and Alberta). Two assessors per farm were responsible for data collection and assessors varied by farms. All assessors underwent an intensive 2-week training program by the same trainer. Standard operating procedures (SOPs) were developed to ensure consistency in measuring and recording the data/information. Periodical checks were conducted by trainers to ensure all assessors remained highly repeatable during the data collection period. An evaluation report was provided and discussed with each producer, identifying strengths and areas for improvement that could benefit dairy cow comfort on their farms. The producers were convinced of the effectiveness of our tool for assessing cow comfort and assisting them to make decisions for improvements. A follow-up study will aim to evaluate the effectiveness of the intervention at initiating changes to improve cow comfort through management and the environment.

Developing Danish welfare indexes for cattle and swine

D.L. Schrøder-Petersen[1], A. Cleveland Nielsen[1], B. Forkman[2], H. Houe[2], H. Feldtstedt[1], L. Lawson[1], J. Tind Sørensen[3], L. Tønner[1], N. Otten[2], S. Norlander Andreasen[2], T.B. Jensen[1] and T. Rousing[3]

[1]*Danish Veterinary and Food Administration, Animal Welfare and Veterinary Medicine Division, Stationsparken 31-33, 2600 Glostrup, Denmark,* [2]*University of Copenhagen, Department of Large Animal Sciences, Grønnegårdsvej 8, 1870 Frederiksberg C., Denmark,* [3]*Aarhus university, Department of Animal Sciences, Blichers Allé 20, 8830 Tjele, Denmark; dlsp@fvst.dk*

In Denmark, a political decision has called for a project with the overall purpose of developing and validating animal welfare indexes at herd level for assessing the development in animal welfare in swine and cattle populations at national level. In addition, indexes at herd level have the potential to assist the animal welfare control system in focusing animal welfare initiatives on problem areas. In the context of the index project, the definition of animal welfare is based on how the individual animal – at the time of observation – experiences its situation, i.e. an affective state definition. The development of the indexes is carried out in a collaborative project involving The Danish Veterinary and Food Administration, University of Copenhagen and Aarhus University. The final models for calculating animal welfare indexes should be ready by 2017. Methodology The project utilizes relevant data available in different national databases – for instance, animal based meat inspection data, age of cattle, race of cattle, etc. and herd level data on mortality, use of antibiotics per animal and the Danish welfare control data. This information – together with additional collected animal and resource-based measurements at herd level – will be used in the development of the welfare indexes. National and international experts are involved in defining and selecting the animal based measurements to be included. A procedure for aggregating measurement data to produce valid and transparent animal welfare indexes is under development. Data for calculating the indexes is expected to be collected by animal welfare inspectors on welfare control visits. Thus collection of animal welfare data should not be time consuming and the indexes should be easy to calculate and the aggregation steps towards the final indexes should be transparent to stakeholders. Before implementation, the indexes will be sought validated in cattle and swine herds using Welfare Quality® protocols as gold standard.

Application of the Welfare Quality® protocols for the evaluation of agricultural policies

A. Bergschmidt[1], C. Renziehausen[1], J. Brinkmann[2] and S. March[2]
[1]Thünen-Institute of Farm Economics, Bundesallee 50, 38116 Braunschweig, Germany,
[2]Thünen-Institute of Organic Farming, Trenthorst 32, 23847 Westerau, Germany;
angela.bergschmidt@ti.bund.de

In the EU's Rural Development Plans (RDP), roughly 300 Mio. Euro have been programmed for the measure 'Animal Welfare Payments' in 2007-2013. While the evaluation of RDPs is mandatory, no commonly agreed methodology exists for the assessment of policy impacts on animal welfare. Against this background, our aim is to use the Welfare Quality® (WQ®) protocols to: (1) evaluate if the Animal Welfare Payments are successful to the effect that animal welfare on the supported farms can be described as 'good'; (2) identify the most successful among the sub-measures of the Animal Welfare Payments; and (3) conclude whether the WQ® protocols have the potential to assess policy effects on animal welfare. In Germany, 'Animal Welfare Payments' are only implemented in a few federal states, among these North-Rhine-Westphalia (~2,000 farms participating) and Mecklenburg-Pomerania (~400 farms participating). For dairy cows, yearly payments per cow range from 37 Euro for straw based housing systems, to 116 Euro for a combination of straw based housing systems, outdoor run and pasture. In a joint effort of two projects, the WQ® protocol is carried out on a random sample of 150 dairy farms participating in the measure in those two federal states. The samples have been stratified according to three most important sub-measures and consist of conventional and organic farms. The survey is carried out from November 2013 until April 2014. In addition to the WQ® protocol, the following information is assembled on the farms: (1) information from the milk control system (i.e. udder health); (2) data (e.g. mortality) from the Identification and Information System for Animals (HIT); (3) resource based indicators for the classification of housing systems (National Assessment Catalogue for Animal Husbandry-NACAH); (4) problem-oriented indicators selected in the partner project. As the collection of data has just started, no empirical results are available yet. Based on the analysis of the farm survey data, our contribution will present results on: (1) Welfare Assessments: – calculation of the welfare categories according WQ® – classification of housing systems according to NACAH – ranking of other indicators. (2) Identification of factors influencing the result of the animal welfare assessment (i.e. effects of the three sub-measures) by means of multivariate analysis. (3) Comparison of the results of the WQ® protocol with the additionally collected indicators/indicator systems.

Towards implementation of farm animal welfare monitoring in practice: example from the veal industry

C.G. Van Reenen[1], M. Wolthuis[1], J. Heeres[1], M. Wesselink[2], M. Cappon[3] and E. Bokkers[3]
[1]Livestock Research, Wageningen University and Research Centre, P.O. Box 65, 8200 AB Lelystad, the Netherlands, [2]Foundation for Quality Guarantee of the Veal Sector, P.O. Box 795, 3700 AT Zeist, the Netherlands, [3]Animal Production Systems Group, Wageningen University, P.O. Box 338, 6700 AH Wageningen, the Netherlands; kees.vanreenen@wur.nl

Between 2005 and 2009, a system for on-farm monitoring of veal calf welfare was developed by a consortium of scientific and commercial partners from France, Italy and the Netherlands. The system comprised of protocols for the recording of primarily animal-based measures of behaviour, clinical health, and post-mortem pathology, and an assessment model based on threshold values for individual welfare measures and the integration of these measures into aggregate scores according to the Welfare Quality® approach. Currently, in the Netherlands, a follow-up study is in progress aiming to support the implementation in the veal industry of a welfare monitoring system. In this study we try to establish a so-called 'quality cycle' whereby (1) the welfare status of a batch of veal calves is assessed by an external, trained assessor, (2) the assessment information is reported and interpreted, and fed back to the farmer by one of his own advisors, (3) remedial steps in terms of e.g. feeding, housing, and management are taken by the farmer if needed and on a voluntary basis; these steps are also laid down in a concise report, and (4) an assessment of the welfare of the next batch of calves takes place. In our project welfare assessments are carried out by inspectors of the Foundation for Quality Guarantee of the Veal Sector, which is an independent inspection body performing food safety and legal compliance audits on Dutch veal farms. Depending on the type of farm and system of veal production, either professional advisors of the veal or feed industry, who already attend veal farms participating in the study, or regular veterinarians act as advisors with regard to the interpretation and possible implications of welfare assessment outcomes. A total of 65 veal farmers voluntarily enrolled in the study. On each farm four consecutive batches of calves will be followed-up and systematically monitored. Wageningen University is responsible for the training and instruction of assessors and advisors, and for data handling and reporting. Throughout the study, we also interview farmers and their advisors, and have feedback meetings with groups of assessors, advisors and farmers. We will address some limitations and advantages of our approach. We suggest that the active involvement of stakeholders within the veal sector in welfare assessment and in providing advice to farmers will facilitate the future adoption of the welfare monitoring system in practice.

Poster 124

Assessing and managing the welfare of animals: various approaches and objectives

B. Mounaix[1], C. Terlouw[2], M. Le Guenic[3], L. Bignon[4], M.C. Meunier-Salaün[5], V. Courboulay[6] and L. Mirabito[1]

[1]Institut de L'Elevage, Service Santé et Bien-être des Ruminants, 149 Rue de Bercy, 75595 Paris, France, [2]INRA, UMRH 1213, INRA de Theix, F-63122 St Genès Champanelle, France, [3]Chambre d'Agriculture de Bretagne, Pôle Herbivores, Avenue Borgnis Debordes, 56009 Vannes cedex, France, [4]ITAVI, UMT BIRD, Centre INRA de Tours, 37380 Nouzilly, France, [5]INRA, UMR 1348 PEGASE, Agrocampus Ouest, 35650 St Gilles, France, [6]IFIP, Bien-être Reproduction, La Motte au Vicomte, 35651 Le Rheu cedex, France; beatrice.mounaix@idele.fr

In Europe, during the last thirty years, regulation was a major drive to improve animal welfare in the different production systems and during transport and slaughtering. In parallel, financial support from industrial stakeholders and political instances have allowed various research actions to meet growing societal expectations. Improvement of animal welfare can be achieved at various levels using more or less sophisticated tools to take into account the multidimensional aspects of welfare. This new, sometimes complex knowledge needs to be transferred to the field in order to improve livestock management from an animal welfare point of view. Therefore, a French network composed of different groups involved in research and development with respect to animal welfare, reviewed main existing welfare assessment and management protocols in pigs, poultry and cattle. The present paper describes the results obtained in cattle. It contains a summary of the main types of welfare indicators in cattle and the description of two complementary approaches illustrated with currently existing tools: evaluation audits for control and certification purposes, and approaches to identify and control risks. The French Charter of Good Practices in Cattle Farming could be one example of such an evaluation audit. Subsequent to this review, the French welfare network intends to identify key questions that need to be addressed when designing certain rearing systems.

Welfare of companion animals in the Czech Republic – evaluation of its supervision

E. Voslarova[1], V. Vecerek[1], J. Dousek[2] and I. Bedanova[1]
[1]University of Veterinary and Pharmaceutical Sciences Brno, Faculty of Veterinary Hygiene and Ecology, Department of Veterinary Public Health and Animal Welfare, Palackeho tr. 1/3, 612 42 Brno, Czech Republic, [2]State Veterinary Administration, Slezska 100/7, 120 56 Prague 2, Czech Republic; voslarovae@vfu.cz

The Czech Republic has been a party to the Council of Europe European Convention for the protection of pet animals since 1998. The text of the Convention was implemented in Czech legislation, specifically in Act No. 246/1992 Coll., on the protection of animals against cruelty, as amended. Regional veterinary administrations are in charge of supervision. The aim of this study was to evaluate the results of supervision over companion animal welfare in the Czech Republic from 1999 to 2012. In the monitored period, the number of inspections and inspected animals and the number of inspections in which deficiencies were detected and the number of animals concerned were observed and the relative frequencies (%) were counted for the individual years. The data were obtained in cooperation with the State Veterinary Administration. The results were analysed using statistic software Unistat 5.1. For the evaluation of a long-term development trend, the actual and relative frequencies for the 1st period (1999-2005) and for the 2nd period (2006-2012) were calculated and frequencies of both periods were compared on the basis of a chi-square analysis of contingency tables 2×2. A total of 52,378 inspections were carried out from 1999 to 2012. Comparing the first and the last seven years of the monitored period, the average relative number of inspections in which welfare failures were detected increased from 7.81 to 11.52% (P=0.0000). The inspections in this field are most frequently carried out upon initiatives of citizens (suspected cruelty to animals or other law violations), and as follow-up inspections following the detection of deficiencies. Thus, the increase of detected welfare failures does not necessary mean a deterioration of companion animal welfare in recent years but may be interpreted as increasing public sensitivity to welfare issues and reporting the possible problems to authorities more frequently. The latter is supported by the increase of annual average number of inspections in the second period.

A sociological analysis of animal welfare regulation in slaughterhouse

F. Hochereau[1] and A. Selmi[2]

[1]INRA, SADAPT, Agroparitech, 16 rue Claude Bernard, 75005 Paris, France, Metropolitan, [2]INRA, SenS, Université paris-est, bois de l'étang, 5bd Descartes; Champ sur marne, 77454 Marne la Vallée, France; adel.selmi@versailles.inra.fr

As part of the implementation of the Regulation 1099/2009, slaughterhouses must prove the effectiveness of killing without pain and establish standard operating procedures (SOP) to comply with animal welfare. Good practice guides have been written up by the professional organisations with some animal welfare organisations. To understand the ins and outs of the regulatory welfare, we examined its implementation to the ground level by different stakeholders (Department of Agriculture, Federation of slaughterhouses, breading and beef institutes, NGOs, consulting veterinarians, official veterinarians) and its application in slaughterhouses. The new 'animal welfare' standard leads to the establishment of a dedicated function at slaughterhouses, the Animal Protection Manager (APM), to the definition of standardised protocols, and to a specific monitoring combined with health and quality requirements. We investigated the beef sector which is more advanced in the application of the regulation. We selected 5 large slaughterhouses (more than 20,000 t), 5 intermediary ones (7,000 to 15,000 t) and 5 small ones (less than 5,000 t) where the APM is the slaughterhouse's manager, the quality manager, the production line manager, or the cowshed manager. Additional Investigations were conducted to assess how animals are handled by workers, and how animal and human sensibility and stress match together. There is a distinction between the symbolic and technical function of APM. The first is often provided by the slaughterhouse top management to ensure ethical slaughtering and traceability of meat quality. The second, provided by middle executives, is focused on the inside of the slaughterhouse. The establishment of SOP varies with the slaughterhouse size: in large structures, SOP are quite already in place whereas in small ones, animal welfare is becoming an organisational challenge. The new regulation imposes training course on animal welfare management and this helped the middle management to define and apply relevant SOP. The animal is rather seen as a barrier to the slaughtering flow pointed out by the number of remaining animals to kill. The faster it is, the faster workers can go home. By this way, animal welfare stands against human well-being. Animal Welfare training will help change this situation. First of all it contributes to a new value put on slaughtering work by qualifying people. It also educates cattlemen on the advantages of proper animal handling according to its perception in order to ease work pressure.

Economic efficiency of small group housing and aviaries for laying hens in Germany

P. Thobe

Thünen Institute of Farm Economics, Institute of Farm Economics, Bundesallee 50, 38116 Braunschweig, Germany; petra.thobe@ti.bund.de

After the ban on cages for laying hens, questions regarding the economic efficiency of keeping laying hens in welfare-friendly alternatives have been raised. Limited information is available on the small group system, which is new and still under debate in Germany. Therefore, an analysis is necessary showing costs, returns and their determinants in small group systems compared with aviaries for laying hens. The calculation of costs and returns is based on a concept proposed by the German Agricultural Society. The empirical basis is a survey of 64 flocks in northern, central and southern Germany presented by a convenient sample. The economic data was collected by an economic expert in a personal meeting after dehousing and then digitized in Microsoft Excel. In both systems the production costs per egg decrease from 13 cents to 6 cents with increasing flock size due to improved physical performance. In small group systems the decline of the returns per egg with increasing flock size is less pronounced than the drop of the costs, so the margin of returns and costs increases. It becomes obvious that farms with larger flocks have economic advantages over farms with small flocks. In contrast to small group systems, the margin of costs and returns in aviaries declines with increasing flock size. This implies that in aviaries economic advantages of larger flocks are less pronounced. The results reflect that good technical equipment and the careful observation of the animals with a 'trained eye' play an important role for a successful egg production. The results show large variations in the economic and physical performance parameters within the size groups. This suggests that in both housing systems the influence of the farm management ranks alongside high physical performance as the most important determinant for cost reduction.

Improving welfare of cod and haddock by adapting current practices in trawl fishing

E.A.M. Bokkers, W.H. Pauw and L.J.L. Veldhuizen

Wageningen University, Animal Production Systems Group, P.O. Box 338, 6700 AH Wageningen, the Netherlands; eddie.bokkers@wur.nl

Fish are sentient beings and therefore animal welfare can be considered an issue in fisheries. Current trawl fishing on cod and haddock leads to exhaustion, injuries and suffocation of the fish, indicators of impaired welfare. Fishermen might be able to implement changes in trawling to improve the welfare of cod and haddock, but they prioritize their economic balance. The aim of this study was to identify improvement options for welfare of cod and haddock during the most welfare impairing phase of catch without economic losses or extra investments to the fishermen. Fish caught in wild capture fisheries have lived a natural life prior to catch. This natural life is interrupted when they are caught and killed by fishermen. In trawl fishing, 5 phases can be distinguished that might affect fish welfare: pretrawl, drifting, towing, surfacing, and landing and processing. Data of a Norwegian trawler was used as a case. Severity of each phase was assessed by studying the contribution to stress and injuries. Also duration of each phase was analysed. In the pretrawl phase (1-12 min), fish swim away from an approaching trawler. During drifting (<2 min), fish try to swim away from the sand cloud that is produced by the fishing gear to lead the fish to the mouth of the net. In these two phases, fish show signs of stress and might be fatigued. During towing (194 min on average), the trawler pulls the net through the water. Increasing crowding in the cod end increases stress, fatigue, and collision with net and other fish, which might result in injuries. During surfacing (15 min), the net is taken up to the water surface leading to compression in the net. This results in oxygen depletion, abrasion and injuries, and differences in temperature, salinity and pressure cause stress. During landing, the fish is exposed to lack of water and oxygen and therefore stressed. When finally processed (after 42 min on average), un-stunned gutting or filleting causes stress and obviously induces injuries. Towing, surfacing, and landing and processing imply the highest severity. Combined with duration, towing is the phase with the highest impact on the welfare of cod and haddock. Shortening towing time and lowering towing speed could improve welfare of cod and haddock in trawl fishing without investments. This, however, might include a decline in total fish catch and thus in revenues for fishermen. Revenues might increase again when higher welfare means higher quality and therefore higher prices. Practical knowledge of fishermen should be deployed to investigate the relationship between product quality and welfare of fish, in order to develop codes of practice for fish welfare in a way that is economically interesting for fishermen.

Harm-benefit analysis – considerations when farm animal species are used in scientific procedures

G.A. Paiba, K. Chandler, K. Garrod and S. Houlton
Home Office, Animals in Science Regulation Unit, 2 Marsham Street, SW1P 4DF London, United Kingdom; giles.paiba@homeoffice.gsi.gov.uk

EU Directive 2010/63 requires that applications for project authorisations should be evaluated and that the evaluation should include a harm-benefit analysis. The evaluation should assess whether the harm to the animals in terms of suffering, pain and distress is justified by the expected outcome taking into account ethical considerations, and may ultimately benefit human beings, animals or the environment. This paper discusses some of the particular issues relating to the use of farm animal species in scientific procedures and how they can affect the harm-benefit evaluation. Factors considered for the harm-benefit evaluation include harms (number of animals being used; species; life stage; overall severity category; specific adverse effects likely to result from the procedures; humane end-points; duration of suffering; method of killing) and benefits (what should be the specific benefits resulting from the project e.g. what will be the scientific outputs; who benefits from the work; how do they benefit; when will the benefits be delivered; what is the likelihood of the benefits being realised) The size and behaviour of many farm species and their particular housing and handling requirements need particular consideration. For example, practicalities and economic considerations mean that group sizes may need to be small and high containment facilities can result in significant constraints on expression of normal behaviours. Male animals may need to be singly housed and female animals may need to be milked. Farm animals are frequently re-used and re-homed. The paper will describe how information provided by the applicant coupled with information gathered during inspection of previous studies along with actual severity reporting and retrospective assessment of projects can all be used to inform the harm-benefit evaluation.

Danish Centre for Animal Welfare

T.B. Jensen, L. Tønner, V.P. Lund and L. Holm Parby
Danish Centre for Animal Welfare, Danish Veterinary and Food Administration, Stationsparken 31, 2600 Glostrup, Denmark; tibj@fvst.dk

The Danish Centre for Animal Welfare (DCAW) was initiated in 2010 and is located at the Danish Veterinary and Food Administration. The centre is a collaboration between animal welfare experts from the Danish Veterinary and Food Administration, Ministry of Food, Agriculture and Fisheries, Aarhus University and University of Copenhagen. The overall aim of DCAW is to contribute to the improvement of animal welfare in Denmark. DCAW contributes to improving animal welfare by presenting an overview of animal welfare-related data from authorities, the farming industry and research, which enables stakeholders to get an overview of the level of animal welfare in various production systems. Hence, a key goal is communicating animal welfare knowledge to relevant stakeholders, such as farmers, politicians, veterinarians, researchers and the general public. DCAW initiates and supports animal welfare research. Since 2010, 28 research projects have been initiated focusing on welfare in both farms and companion animals. Research themes have so far included: (1) using register data to evaluate animal welfare at herd level; (2) methods to measure animal welfare in farms; (3) the effect of transportation on the welfare of animals; (4) economic consequences of improving animal welfare in farms; (5) problems in relation to sick and injured animals; (6) animal welfare in relation to companion animals; and (7) teaching and communicating animal welfare issues. Results from the initiated projects are each year presented at an annual conference hosted by DCAW. At the conference new developments and findings relevant to animal welfare issues are also presented. Finally, a main priority is communicating and collaborating with European and non-European countries. DCAW wishes to share information and ideas about the national control of animal welfare as well as new animal welfare initiatives. An international collaboration will provide a common contribution to the improvement of animal welfare.

Farm animal welfare reporting in Germany: what can we learn from our neighbours?

S. Starosta and A. Bergschmidt
Thünen-Institute, Farm Economics, Bundesallee 50, 38116 Braunschweig, Germany;
sonia.starosta@ti.bund.de

Currently, farm animal welfare is an important and politically sensitive topic in German society and the EU. The public expresses their strong interest via various animal welfare initiatives, calling for higher welfare standards. On the other side, a majority of livestock farmers argue that the welfare situation of their animals is uncritical. Despite these conflicting perspectives in society, there is little neutral information available on animal welfare in Germany. The goal of this study is to examine European reports dealing with animal welfare to identify lessons learnt for the structure and content of a German animal welfare report. Publications by official institutions have been selected for the assessment, as the intended German report will be published by a governmental agency. In a first step, an internet research was carried out (languages: English, French, German). This yielded reports from Finland (FIN), Denmark (DK), Switzerland (CH), United Kingdom (UK) and Austria (AT). Additionally a request for animal welfare reports was sent to the 28 Ministries for Agriculture in the EU; as only UK, DK and AT responded, no additional information was generated. Though animal welfare reports had been searched for, only accounts on animal protection could be obtained and examined. The main difference between the two types of reports is that animal protection reports inform about the actions undertaken to protect animals while animal welfare reports cover the welfare of the animals. Nevertheless, animal protection reports may contain relevant information on animal welfare too. The enforcement of legislation through frequent, numerous inspections and dissuasive sanctions are a precondition for minimum welfare standards in livestock husbandry, transportation and slaughter. Additionally to the legal setting, reports might contain animal welfare related information. As can be expected from animal protection reports, the five examined sources concentrate on infringements. AT, FIN and DK give detailed information on the kind of infringements that occurred e.g. missing illumination in hen houses, exceeded thresholds of dermatitis incidences in broiler flocks; all without mentioning the number of affected animals. Next to the infringements AT, FIN and DK give insights into imposed punishments and affected animal species. Only FIN delivers some animal related data, e.g. the amount of disqualified slaughter carcasses as an approximate indication for sick or injured animals, as well as the duration of transports (amount of loads <9 and >9 hours). Still, none of the examined reports can serve as a concrete model for a German blueprint as animal welfare aspects were not or only vaguely touched.

The role of trusted individuals in the effective communication of animal health and welfare research

J. Hockenhull[1], G. Olmos[1], L. Whatford[1], H.R. Whay[1], D.C.J. Main[1], S. Roderick[2] and H. Buller[3]
[1]University of Bristol, School of Veterinary Sciences, BS40 5DU, United Kingdom, [2]Duchy College, Rural Business School, TR14 0AB, United Kingdom, [3]University of Exeter, Geography, College of Life and Environmental Science, EX4 4RJ, United Kingdom; jo.hockenhull@bristol.ac.uk

The impact of advances in animal health and welfare research relies on findings being disseminated to those who can implement changes on a practical level. This dissemination must be done in such a way that people are motivated to alter their current practices. The role that trusted individuals, e.g. industry professionals, veterinary surgeons and farmers, can play in this process could be significant but is often overlooked. The EU funded South West Healthy Livestock Initiative (SWHLI), provided an opportunity to explore the effectiveness of various communication and animal health awareness-enhancing strategies directed at those farming in the South West of England. SWHLI was comprised of ten independently run sub-projects which utilised a range of methodologies and communication styles, all of which had the objective of raising awareness of key health issues in the south west livestock sector. Each project relied on the involvement of veterinary surgeons, to varying extents, for the recruitment of participants and information delivery. In the final year of the initiative, semi-structured interviews were conducted with 51 key stakeholders associated with sub-projects encompassing the beef, dairy, sheep, broiler and laying hen sectors. The aim was to explore the general perception of the projects, levels of engagement with them and their impact in terms of increased awareness and/or changes in practice. Participating farmers spoke of the important role of trusted individuals in two distinct contexts. First, individual relationships with a farmer's own familiar veterinary surgeon played a key role in motivating engagement with the project. It is argued that personal trust of their vet was more important to many than the nature of the information being communicated about animal health through the schemes. Second, the project also identified the key role played by respected farming colleagues in sharing practical experiences in dealing with animal health and disease, through organised talks or more beneficially through farm walks, enabling the practicalities of on-farm changes to be discussed and potential, real-life solutions suggested. The emphasis placed on trust and respect of key individuals in stimulating engagement and on-farm changes was pronounced and suggests that such individuals are a valuable resource in the effective communication of animal health and welfare research findings to the farmers, not least because of the community that they generate around them.

Significance and market positioning of animal welfare products in the German food retail sector

W. Pirsich, H. Heise and L. Theuvsen

Georg-August University of Goettingen, Department of Agricultural Economics and Rural Development, Platz der Goettinger Sieben 5, 37073 Goettingen, Germany; wpirsic@gwdg.de

Ethical and sustainable aspects gain increasing importance when purchasing food products. In this context food labels can serve consumers as an important guide. Recently, the establishment of products with animal welfare labels can be observed in Germany. These food labels are an attempt to create a market segment for products of animal origin, in which consumers are willing to pay higher prices for the compliance with higher animal welfare standards. Numerous consumer studies have shown a willingness to pay 10-35% more for such products. Whereas consumers have been surveyed frequently, studies on the significance of animal welfare friendly products for retailers are rare so far. This paper provides an overview of the distribution structures of products with animal welfare labels in the self-service range of German food retail. Through a store check, a data collection was carried out in discount stores, supermarkets and hypermarkets in Hamburg, a metropolitan area in Northern Germany. Information on the availability of animal welfare products on the homepages of the retailers served as a starting point for data collection. Type and quantities of the products offered, prices, placement and marketing activities were taken into account. The results show that in the most retail stores, pigs and/or poultry products with animal welfare labels are offered. We found seven different labels with reference to higher animal welfare standards. Overall, the product range turned out to be narrow and prices vary significantly between the various labels. Specific marketing activities for these products could only be observed in a few cases. The results of the store check were validated through expert interviews with store and procurement managers of the regional EDEKA retail company. EDEKA is a cooperation of independent retailers and the market leader in Germany. Nationwide more than 4,000 independent EDEKA merchants decide on their product range on their own responsibility. In each case the product range is tailored to the individual needs of the local customers. The interviews focused on the experts' assessment of the significance and perspectives of animal welfare products in the retail sector. We conducted semi-standardized face-to-face interviews as well as a questionnaire-based survey of store managers. First analyses indicate that the food retail sector is very much interested in animal welfare issues. Food retailers see several reasons for the very low market penetration of animal welfare friendly products, for instance the need to develop new marketing concepts in order to take advantage of the frequently described market potential.

Pig producer perspectives on the potential of meat inspection as an animal welfare diagnostic tool

L.A. Boyle[1], C. Devitt[2], A. Hanlon[2] and N.E. O'Connell[3]
[1]Teagasc, Pig Development Department, Animal & Grassland Research & Innovation Centre, Moorepark, Fermoy, Co. Cork, Ireland, Ireland, [2]University College Dublin, Veterinary Sciences Centre, University College Dublin, Ireland, Belfield, Dublin 4, Ireland, [3]Queens University, Institute for Global Food Security, School of Biological Sciences, Northern Ireland Technology Centre, Belfast, Northern Ireland, United Kingdom; laura.boyle@teagasc.ie

This work aims to develop pig welfare lesions recorded at ante and post (carcass) mortem meat inspection (MI) as a welfare diagnostic tool. Social science research was employed to establish stakeholder perspectives towards this development. This paper presents results from semi-structured telephone interviews conducted with 18 pig producers. Interview data were inputted into NVIVO 8 and thematic network analysis was used for data analysis. The COREQ-32 (Consolidated criteria for reporting qualitative research) checklist was used to ensure quality control in methodology, analysis and reporting. Results reveal producers levels of satisfaction with current information received from the slaughterhouse, their expectations concerning the role and responsibility of those involved in MI, beliefs about on-farm health and welfare problems, and producer aspirations for the potential use of MI as a welfare diagnostic tool. Results highlight the centrality for producers, of the private veterinary practitioners in on-farm pig health and welfare. The potential use of MI as an animal welfare diagnostic tool is undermined by issues of trust and confidence among pig producers in the MI process and the extent to which information would be used for the benefit of the producer. Producer tolerance and acceptability of welfare issues, such as tail-biting also undermines the potential of MI as a diagnostic tool. Improved, consistent information on the health and welfare of their pigs would help build trust across the supply chain and enable producers to identify trends – empowering them to recognise the impacts of health and welfare issues on pig production.

PigSurfer – SURveillance, FEedback & Reporting within ProPIG for communication with 75 pig farmers

C. Leeb[1], D. Bochicchio[2], G. Butler[3], S. Edwards[3], B. Frueh[4], G. Illmann[5], A. Prunier[6], T. Rousing[7], G. Rudolph[1] and S. Dippel[8]
[1]Univ. of Natural Resources and Appl. Life Sciences (BOKU), Dep. of Sustainable Agric. Systems, Gregor Mendelstr. 33, 1180 Vienna, Austria, [2]CRA-SUI, Italy, Agricultural Research Council, San Cesario sul Panaro, 41018 Modena, Italy, [3]School of Agriculture, Food & Rural Development, Newcastle University, Agriculture Building, Newcastle upon Tyne NE1 7RU, United Kingdom, [4]FiBL, Animal Husbandry Division, Ackerstrasse 21, 5070 Frick, Switzerland, [5]Institute of Animal Science, Pratelstvi 815, 10400 Prague, Czech Republic, [6]INRA, UMR1348 PEGASE, 35590 Saint-Gilles, France, [7]Aarhus University, Dept. of Animal Science, Blichers Allé 20, 8830 Tjele, Denmark, [8]Friedrich-Loeffler-Institut, Institute of Animal Welfare and Animal Husbandry, Doernbergstr. 25/27, 29223 Celle, Germany; christine.leeb@boku.ac.at

The CoreOrganic2 research project ProPIG is carried out in 75 organic pig farms in 8 European countries (AT; CH; CZ; DE; DK; FR; IT; UK) to improve animal health, welfare and nutrition using farm customised strategies. For future on-farm application (e.g. advisory/certification activities which are carried out during one day visits), an automatic recording and feedback tool was developed. This should allow on-farm data collection, import of data into a database and the possibility for benchmarking, including a printed output for the farmer to facilitate immediate discussion of results and improvement strategies. To document not only animal health and welfare, but also integrate diet composition and productivity data, it was important to choose key indicators from all areas, which would be available across all eight countries. Based on existing on-farm welfare assessment protocols (e.g. WelfareQuality®, CorePIG) indicators were selected by the consortium, which were then transferred into a Software programme ('PigSurfer'). This Software is available as Desktop- or Android version to be used on Tablet PCs, so that on-farm data (interview with farmer; direct observations on weaners, finishers, sows; productivity and treatment records; feed) can be entered directly. During two visits a database was built and a 'Farm report' was printed for each farm including benchmarking of results for feedback and discussion with the farmer. After a year 'PigSurfer' was used to carry out the following complete process during one day visits across Europe: Surveillance of health and welfare, feedback of data in comparison with results from the previous year as well as benchmarking with 75 other pig farms and printing a report. 'PigSurfer' is a promising tool for communicating health and welfare, as it provides not only a database, which can be continuously extended, but is an important step to move from research to on-farm application across Europe.

Poster 136

A survey of management practices that influence production and welfare of dairy calves in 244 family

M.J. Hötzel, C. Longo, L. Balcão, J.H.C. Costa, C.S. Cardoso and R.R. Daros
Universidade Federal de Santa Catarina, LETA, Laboratório de Etologia Aplicada e Bem-Estar Animal, Rod. Admar Gonzaga, 1346, Itacorubi, 88034-001, Florianópolis, SC, Brazil; mjhotzel@cca.ufsc.br

The aim of this paper is to report dairy calf management practices used by 244 smallholder family farmers in the South of Brazil. Data were collected via a semi-structured questionnaire with farmers, inspection of the production environment and an in-depth interview with a sample of the farmers to understand the reasons for their choices of management practices. Herds had an average of 22.3 (range 5-86) milking cows and average milk production was 12.8 (range 4-31) l/cow/day. The main breed of the herds was Jersey (53%), Holstein (38.8%), and undefined breed (8.2%). Calves were separated from the dam up to 12 h after birth in 78% of the farms, and left to nurse colostrum from the dam without intervention in 54% of the farms. The typical amount of milk fed to calves was up to 4 (range 2-7) l/day, in two meals, until a median age of 75 days. In 40% of the farms milk was provided in a bucket, in 49% with teats, and in 11% calves suckled from a cow. Preweaned calves were housed individually in 70% of the farms, mostly in indoor stalls (81%); 13% of the calves were kept on pasture, in groups. Calves were dehorned in with a hot iron, and male calves were surgically castrated in 95% and 79% of the farm, respectively; no pain control was used for these interventions. Farmers reported that the male calves were killed on the farm shortly after birth in 35% of the farms; 51% reported that calves were reared for consumption, sold or donated to other farmers and 14% that some calves were killed and some were reared, according to resource availability. Diarrhoea was reported as the main cause of calf mortality in 71% of the farms, though mortality was assessed as low by farmers. No records on calf mortality, disease or use of medicines were available in the farms visited to support these reports. Perceptions related to exceeding labour demand seemed a key element underpinning the low adoption of innovations that may improve the quality of dairy calves' care. Changing this scenario is essential to support the sustainable development of dairy production, an activity of great economic and social relevance for the region. To that end farmers need to be informed on the production, animal welfare and economic outcomes of calf feeding, neonatal care, housing, and the need to keep records to help base decisions for management choices. Also needed is the development and promotion of opportunities for ethical, economic uses of male calves, and practical ways to dehorn and castrate calves on the farm with scientifically validated methods to minimize pain and suffering.

The animal welfare science hub: An open content management system that promotes welfare assessment

A. De Paula Vieira[1], F. Langford[2], J.B. Vas[3], P. Gomes[1], D.M. Broom[4], B. Braastad[3] and A.J. Zanella[5]
[1]Universidade Positivo / AWIN Project, Design, R. Prof. Pedro Viriato Parigot de Souza 5300, Bloco Vermelho, Curitiba/PR, 81280330, Brazil, [2]SRUC / AWIN Project, Animal & Veterinary Sciences, SRUC, Roslin Institute Building, Easter Bush, Midlothian, EH25 9RG, United Kingdom, [3]Norges Miljø- og Biovitenskapelige Universitet, Postboks 5003, 1432 Ås, Norway, [4]University of Cambridge, Madingley Road, Cambridge CB3 0ES, United Kingdom, [5]Universidade de São Paulo, Departamento de Medicina Veterinária Preventiva e Saúde Animal, Av Duque de Caxias Norte, 225, 13635-900, Brazil; apvieirabr@gmail.com

The Animal Welfare Indicators Project (AWIN) launched the Animal Welfare Science Hub (www.animalwelfarehub.com), an open content management system that allows for collaborative sharing of materials that are useful for animal welfare teaching, learning, outreach, animal welfare assessment and research purposes. The Hub provides a platform to update users on the latest animal welfare news, advertise animal welfare assessment courses, provide training via webinars and learning objects on the latest welfare assessment methods and help users to solve animal welfare problems in a collaborative and interactive 3D environment. AWIN is seeking content to be hosted in the Hub, e.g. courses, videos, animations, mobile Apps and 3-D simulations, which will need to be portable and have good usability, applicability and teaching quality. All peer-reviewed materials are accessible to instructors in the classroom or online, by students and other interested parties worldwide. The Hub also allows users to initiate discussions and share their views. In addition, Hub users can consult a research database on educational resources and on the latest published animal welfare research, create online surveys, share their favourite animal photographs, find out about upcoming events on animal welfare around the world, find jobs and funding opportunities in their areas of expertise and download mobile Apps. At this conference, AWIN will also demonstrate the first Apps that were developed to train horse owners and veterinarians to assess pain in horses and conduct welfare assessment in turkeys. All Hub teaching materials go through a process of external review by the International Society of Applied Ethology (ISAE) to ensure that everything presented is scientifically valid, educationally effective and user-friendly.

Poster 138

Is the maternal behaviour conditioned by emotional reactivity in Yucatan sows?

J. Jansens[1], D. Val-Laillet[2], C. Tallet[1], C. Houdelier[3], C. Guérin[1], S. Lumineau[3] and M.C. Meunier-Salaün[1]
[1]INRA-AgroCampus Ouest, UMR1348 PEGASE, UMR1348 PEGASE, 35590 Saint-Gilles, France, [2]INRA ADNC UR1341, ADNC UR1341, ADNC UR1341, 35590 Saint-Gilles, France, [3]Université Rennes I, UMR 6552 Ethos, Campus Beaulieu, 35042 Rennes, France; marie-christine.salaun@rennes.inra.fr

Genetic selection on sows is usually focused on performance-based criteria, such as prolificacy or milk production. Though, criteria linked to the maternal behaviour per se are more considered for increasing piglet and sow welfare. Moreover, the link between the female's emotional reactivity and its maternal behaviour, reported in rodents and poultry, has received little attention in pigs. An experiment was conducted in order to evaluate this relationship in Yucatan sows, an experimental model for behaviour, nutrition and neurosciences studies. Eight two-year-old, multiparous (2-3) and loose-housed Yucatan sows were studied. The emotional reactivity was determined in standardised tests: open-field (OF) and Human in an arena test 18 d prepartum, and novel object test in the housing pen 10 d prepartum. Maternal behaviour was analysed by 10-min scan sampling for the 24 h prepartum, by continuous sampling during the farrowing process and for 24 h at 3, 5, 11, 15 and 32 days postpartum. Principal component analysis (PCA) on emotional reactivity data allowed us to identify 2 groups with differentiated behavioural patterns. Group 1 was mainly characterized by higher exploration of the OF pen, the unknown human and the novel object, while Group 2 was mainly characterized by more time spent immobile, a higher distance from the human, and higher latency to enter the OF arena and investigate the human or object. A comparative analysis between both groups showed some differences in maternal behaviour, but significant only during the postpartum period, time to investigate the pen higher in group 1 (U=16 n1=n2=4, P<0.05). Correlations were also found between the pen investigation during the OF test and the piglet investigation by the sow during farrowing (r=-0.69, P=0.06). The latency to investigate the unknown human was correlated to the number of times the sows was observed pushing their piglets during the farrowing (n=8, r=0.74, P<0.05) and the suckling frequency during postpartum (r=0.68, P=0.06). The time spent close to the unknown human was correlated with the frequency of interrupted suckling by the sows (r=0.84, P<0.01). In conclusion, sows displaying higher fear or anxiety responses in the standardized tests used in this study to assess emotional reactivity, tended to show higher maternal abilities. Those results need to be confirmed but highlight the importance of the assessment of the emotional reactivity for the prediction of maternal abilities, and behavioural criteria or markers for temperament-based selection.

Poster 139

Sows postural changes, responsiveness towards piglets screams and theirimpact on piglet mortality

G. Illmann, M. Melišová and H. Chaloupková
Institute of Animal Science, Department of ethology, Pratelstvi 815, 10400, Czech Republic; gudrun.illlmann@vuzv.cz

Free farrowing pens (pens) improve the welfare of sows, but might increase sow activity and could negatively influence piglet production. The objective of this study was to assess the effect of pens and crates on sow postural changes, piglet trapping, sow responses towards piglet screams, piglet mortality, and piglet body weight (BW) gain. It was predicted that provision of increased space (pens) would increase the frequency of sow postural changes and the probability of trapping, but also sow responses towards the screams of piglets; thus there would be no differences in fatal piglet crushing or overall mortality between either housing system. Sows were randomly moved to a farrowing pen (n=20) and farrowing crate (n=18) placed in the same room. Sow behaviour was recorded and analysed for 72 h from the birth of the first piglet (BFP). Sow posture changes included: rolling from a ventral-to-lateral position and vise versa, standing-to-sitting, standing-to-lying, and sitting-to-lying. Occurrences of piglet trapping and sow responsiveness towards real crushing situations were analysed. Sow responsiveness was assessed towards audio playbacks (PB) with piglet screams on day 3 postpartum (48-72 h after BFP; PB Crush Calls) and real piglet crushing during the first 72 h after BFP (Real Crush Calls). Piglet BW gain was estimated 24 h after BFP; piglet BW at weaning; piglet crushing and piglet mortality during the 72 h after BFP. The data were analysed using PROC MIXED and PROC GENMOD. Sows in pens showed more postural changes (P<0.05) and tended to have greater incidences of piglet trapping than those in crates. Sow response towards PB Crush Calls was greater in pens (P<0.05), but did not differ from Real Crush Calls between pens and crates (NS). There was no significant effect detected on the probability of piglet crushing (NS) and mortality (NS) during the 72 h after BFP, nor in piglet mortality at weaning (NS) between pens and crates. Piglet BW gain at 24 h after BFP (P<0.05) and piglet BW at weaning (P<0.05) was significantly greater in pens. In conclusion, as predicted, sows in pens showed more postural changes and tended to trap more piglets; however, the response towards Real Crush Calls did not differ between the two housing systems. Despite this, there was no increase in piglet crushing or piglet mortality in pens, which might be influenced by the better body condition of piglets observed in pens, which in turn could influence their ability to avoid crushing by the sow.

Behavioural assessment of livestock guarding dogs

B. Ducreux

Institut de l'Elevage, 149 Rue de Bercy, 75012 Paris, France; barbara.ducreux@idele.fr

When at the end of the 19th century wolves, lynx and bears were eradicated in France, new forms of livestock, which can provide added-value to natural handicaps areas, blocked off lands where these predators lived. For around twenty years, the natural or resulting of releases return of these predators has unfold in the mountains (Alps, Jura and Pyrenees). In these lands, flocks are subjected to attacks which nurse passionate debates between farmers and wildlife supporters. To promote coexistence between predators and pastoralism, the state set up a special aid program to help farmers to finance the implementation of protective measures. Livestock guarding dog is one of the existing tools, the only which is responsive and is capable of getting used to existing diversity of predators. The advantage is that dogs behave independently, without having need of orders to take action; it may be a problem because they work at the heart of multi-use areas in which men live. To make sure that the livestock guarding dog, who works in a land, can ensure optimal protection of the herd without posing a risk to humans, a behavioural assessment test was built at the request of Ministry of Agriculture and livestock guarding dogs' users. The long-term goal is to reduce heterogeneity that exists on the ground by making a selection and a qualification of breeding livestock guarding dogs. This test was formalized by scientists (ethologists and veterinarians) and experts (dog's testers). The first step was to determine the essential qualities that an effective livestock guarding dog has to show: his attachment to the herd, his loyalty and his attentions. Then, to assess these qualities, best variables to observe and quantify during trials where men act have been chosen. Finally, the test is composed of six trials which allow assessing livestock guarding dog's behaviour in relation to his herd, his reactions in front of unknown human and in front of unusual, surprising or unsettling element (bike passing and gunshot). For each trial, two people work together: one makes practical tests and the other reports information about dog behaviour. For safety, it is essential that two people remain close to each other throughout the test. This test will be deployed and used by empowered testers in the coming months. If they are relevant and allow a better assessment of the dog, some minor changes may be made.

Dairy calf characteristics and the association with performance later in life

E.A.M. Bokkers[1], A. Coysman[1] and C.G. Van Reenen[1,2]
[1]Wageningen University, Animal Production Systems Group, P.O. Box 338, 6700 AH Wageningen, the Netherlands, [2]Wageningen University & Research Centre, Livestock Research, P.O. Box 65, 8200 AB Lelystad, the Netherlands; eddie.bokkers@wur.nl

Early life characteristics of calves may have the potential to improve the welfare of dairy cattle when they are able to predict cow performance in terms of fertility, production, and health later in life (lifelong performance). This might give the opportunity to early select promising animals for longevity, which is a sign of good welfare but also reduces economic costs (less young stock needed) and therefore contributes to a sustainable dairy production. This study aimed to investigate the performance of cows at three life stages (calf, heifer, cow) and to associate life stage performance with lifelong performance. A dataset was used including measures on activity, behaviour and physiological responsiveness to environmental challenge of 374 calves, 336 heifers and 688 cows (all Holstein Friesian). In total, 842 unique animals were included in the study. From 92 animals data was available of all three life stages. Within each life stage a principle component analysis was performed to reduce the number of observed variables into a smaller set of principle components (PCs), explaining the largest percentage of variation of the original dataset. Spearman rank correlations were calculated between PCs and measures of performance (production, growth, health and fertility), between PCs obtained in different life stages, and between measures of performance within and between life stages. Several significant correlations between PCs per life stage and measures of production, growth, health and fertility were found, but all had low values ($rs \leq 0.18$). Correlating the PCs between life stages, only resulted in a significant correlation between the PCs 'Cow fear' (based on milkability and fear score) and 'Heifer regularity' (consistent day to day activity patterns) ($rs=-0.23$) and between the PCs 'Cow regularity' (consistent day to day activity patterns and number of times lying down) and 'Heifer eating' (latency to eat after transport)($rs=-0.22$). Several significant correlations between production, growth, health and fertility variables were found, but again all correlations were low ($rs<0.4$). Some examples: birth weight was correlated with the number of vital ova/animal/yr ($rs=0.26$) and with protein content of the milk ($rs=-0.20$). Also, the number of medical treatments after calving was correlated with the number of heats/cow/yr ($rs=-0.35$). In conclusion, this study found significant but low correlations between life stage- and lifelong performance. The practical use of a set of early life indicators, as a tool for the selection of promising replacement stock, remains to be demonstrated.

The use of hospital pens in Danish dairy production

C. Amdi[1], M.B. Jensen[2], M.S. Herskin[2], P.T. Thomsen[2], B. Forkman[1] and H. Houe[1]
[1]Faculty of Health and Medical Sciences, University of Copenhagen, Department of Large Animal Sciences, Grønnegårdsvej 2, 1870 Frederiksberg C, Denmark, [2]Aarhus University, Department of Animal Science, Blichers Allé 20, 8830 Tjele, Denmark; houe@sund.ku.dk

The provision of comfortable housing for sick and injured animals is an animal welfare issue. Danish legislation states minimum space requirements for hospital pens and that the bedding must be soft. In addition, it will be a legal requirement that Danish dairy herds must have at least one hospital pen place per 100 cows in the herd from 2016. However, little is known about how Danish milk producers use hospital pens in the treatment of sick cows or about the present type (individual or group pen), size and number of hospital pens available on farms. The aim of this study was to provide background information on the current use of hospital pens in Danish dairy herds. Three-hundred and fifty randomly selected cattle farmers were given a questionnaire on the use of hospital pens on their farm. One hundred and thirty-five farmers (39%) responded, of which 86 kept dairy cows on their farm. Farmers were asked questions about their production system, the number and characteristics of hospital pens available and how the cows were managed while in the hospital pen. Among the 86 replies, 16% had access to group hospital pens, 30% had single cow hospital pens, 46% had both types of hospital pens, and the remaining 8% did not have access to hospital pens at all. This is to our knowledge the first time it has been investigated how and to what extent farmers use hospital pens in Danish dairy herds.

Variation in health of dairy calves when uing automated feeders in the USA

M. Jorgensen[1], A. Adams-Progar[1], A.M. De Passille[2], J. Rushen[2], S. Godden[3] and M. Endres[1]
[1]University of Minnesota, Department of Animal Science, 1364 Eckles Avenue, St. Paul, MN 55108, USA, [2]University of British Columbia, 2357 Main Mall, Vancouver, BC V6T 1Z4, Canada, [3]University of Minnesota, Veterinary Population Medicince, 1365 Gortner Avenue, St. Paul, MN 55108, USA; jorge505@umn.edu

Research is very limited regarding best housing and management practices for automated calf feeding systems, particularly in terms of how these factors influence animal welfare. This ongoing study is characterizing health scores, morbidity and mortality of group-housed calves in US farms and relating these to housing and management practices. Thirty-eight dairy farms in the Midwest of the USA were visited every 60 days. During each visit, calves (n=7779) were scored for health using four categories: attitude (0-4); ears (0-4); nose (0-3); eyes (0-3); and cleanliness (0-2), with 0 representing a normal, healthy calf. During each season, milk samples were collected from the mixing container inside the feeder and the tube leading to the nipple for measurement of standard plate count (SPC) and coliform count. Pearson's correlation coefficient was used to analyse the relationship between mean SPC and health scores. There was a large variation between farms in calf health. On the 10 farms with the best health scores, a mean of 9.7% (range of 2.9-12.9) of animals displayed abnormal scores for attitude, 3.7% (1.7-5.1) for ears, 12.2% (7.8-14.8) for nose, 7.2% (2.0-11.9) for eyes, and 26.4% (20.1-32.6) for cleanliness (an indicator of diarrhoea). On the 10 farms with the worst health scores, a mean of 22.8% (15.7-30.3) of animals displayed abnormal scores for attitude, 14.4% (10.0-22.5) for ears, 27.2% (22.8-30.6) for nose, 30.3% (22.5-36.4) for eyes, and 54.9% (50.6-60.3) for cleanliness. SPC (cfu/ml) in the mixing tank varied between seasons, being lowest in spring (median=3,900; Q1=860, Q3=252,500) and highest in summer (median=32,250; Q1=2,395, Q3=346,250). SPC in the tube also varied between seasons, being highest in winter (median=510,000; Q1=137,500, Q3=3,450,000) and lowest in autumn (median=185,000; Q1=35,000; Q3=5,250,000). No relationship was observed between tube SPC and attitude, ears, nose, or eyes scores; however, SPC was significantly correlated with calf cleanliness scores (r=0.26, P=0.002). The variation in health scores among farms shows that welfare in automated feeder systems can be improved. In addition, results indicate that the cleanliness of automated feeder equipment may influence calf health; however, further data collection and analyses of calf morbidity and mortality should provide a more complete understanding of risk factors.

Poster 144

Behaviour and animal welfare according to lairage time in Uruguay

F. Costa[1] and M. Del Campo[2]
[1]*UNESP, Zootecnia, Via de Acesso Prof. Paulo Donato Castellane s/n, CEP: 14884-900 Jaboticabal, SP, Brazil,* [2]*INIA, Ruta 5 Km 386, Tacuarembó, Uruguay; mdelcampo@tb.inia.org.uy*

Uruguay exports 80% of its meat production. In this context, is necessary to consider the intrinsic value of meat, as well as the ethical quality of the processes involved. The aim of this study was to evaluate the effect of pre slaughter lairage time on behaviour and animal welfare through a multidisciplinary approach. Thirty Hereford steers were randomly grouped in two treatments according to lairage time in slaughterhouse pens: 12 hours (Treatment 1, n=15) and three hours (Treatment 2, n=15). Animals were slaughtered when they reached an average of 500 kg of live weight in a commercial abattoir licensed to export meat, and following animal welfare standard procedures. The behavioural records of the animals were performed during lairage by direct observation and were obtained through the identification of body posture (standing), the record of rumination activity and the frequency of negative behaviours (fighting and mounting). Blood samples were taken from all animals looking for basal values in welfare indicators such as Creatine kinase (CPK) and Glucose and their respective changes according to the following periods: at the farm, after transport, after lairage, and during bleeding. The journey lasted 5 hours, with a truck in good condition and a proper driving. GLM procedures were used to analyse behaviour, and Glimmix for repeated measures were used to study major effects on blood indicators through time. and All animals were standing during the first 3 hours in lairage pens and negative behaviour showed the highest frequency during the first 3 hours in both Treatments. Rumination frequency was high until 9 hours in Treatment 1, suggesting no hungry problems with fasting time in this Treatment (12 hours of lairage). Animals from Treatment 2 (3 hours) increased CPK concentration after lairage, possibly related to the high frequency of mount and fighting observed during this period. In addition, animals from Treatment 2 had higher glucose concentration during bleeding. Both physiological indicators, suggested that these animals did not have enough time to recover from transport. In conclusion, the first 3 hours in the slaughterhouse are critical for animal adaptation. Steers that remained in pens could rest and recover afterwards, being this not possible for Treatment 2. Results from this experiment with animals coming from extensive conditions and with short time of transportation, suggested that animal welfare should be less compromised with the longer lairage.

Measuring stress level of dairy cows when changing from conventional to robotic milking

V. Jurkovich[1], M. Kulcsár[1], L. Kézér[2], J. Tőzsér[2] and L. Kovács[1,2]

[1]*Szent István University, Faculty of Veterinary Science, István utca 2, 1078 Budapest, Hungary,* [2]*Szent István University, Faculty Agricultural and Environmental Sciences, Páter K. u. 1, 2100 Gödöllő, Hungary; jurkovich.viktor@aotk.szie.hu*

In intensive dairy farming, the milking technology is a main factor in determining the welfare of animals. As adapting to environmental changes can be challenging for high-producing cows, the readjustment of the milking system from conventional to automatic can cause stress to animals. Our aim was to measure the changes in the stress level of cows when the milking technology changed from conventional to robotic milking on a particular farm. In the first of the two experimental periods, all cows had been milked in a conventional herringbone milking parlour, while in the second period all cows were milked in a robotic milking unit (DeLaval VMS). The heart rate (HR) and heart rate variability (HRV) parameters of dairy cows (n=18; parity: 2.3±1.1; milk yield: 24.4±6.8 kg/d; DIM: 148.3±57.4) during milking in a conventional system were compared to those of the same animals when milked in an automatic milking system. HRV parameters were analysed in frequency- (LF, HF, LF/HF) and geometric domain by Poincaré plot (SD1, SD2, SD2/SD1). Faecal samples were also taken from the cows in the morning and in the evening for cortisol concentration. The samples were analysed by RIA. When conventional milking was in operation, HR was higher in the holding area, before entering the milking parlour/milking unit, during main milking, during the last minute of milking, during the entire milking and during the whole time spent in the milking stall/milking unit (P<0.001; P<0.001; P=0.010; P=0.028; P=0.046; P=0.005, respectively) compared to the automatic milking system. Sympathetic HRV parameters SD2 and SD2/SD1 measured in the conventional system during the phases of waiting in the holding area, before entering the milking parlour/milking unit, and during udder preparation were also higher compared to automatic milking (P=0.012, and P=0.013; P=0.0012 and P=0.004; P=0.033, and P=0.031, respectively). The average daily faecal cortisol concentrations were higher (15.1 vs 5.6 nmol/l, P<0.01) when the conventional milking process was in operation. Based on our results, the conventional milking process imposed greater stress on the animals. The cortisol concentrations showed a higher stress level in general, and on the basis of HR and HRV analysis the biggest differences occurred in the holding area, before entering the milking parlour/milking unit and during udder preparation.

Slaughtering of non-castrated boars and application of boar taint detection methods in Belgium

J. Vicca[1], F. Buysse[1], A. Van Den Broeke[2] and M. Aluwé[2]
[1]HUBKAHO, Agro- and biotechnology, Hospitaalstraat 23, 9100 Sint-Niklaas, Belgium, [2]Institute for Agricultural and Fisheries Research (ILVO), Animal Sciences, Scheldeweg 68, 9090 Melle, Belgium; jo.vicca@hubkaho.be

In preparation to the European ban on surgical castration of male piglets on 1 January 2018, Belgian slaughterhouses were questioned about their current use of boar taint detection methods and their future perspectives. Slaughterhouses were selected on basis of their annual number of slaughtered pigs, which should be at least 10 200. This criterion was fulfilled by 19 out of 38 pork slaughterhouses in Belgium. Ten of them were interested to participate. Questionnaires finally were filled in by 8 slaughterhouses. Following topics were asked for: (1) number of intact / immunocastrated boars slaughtered each week; (2) presence of a boar taint detection system + its characteristics; (3) requirements for a boar taint detection system; (4) destination of carcasses with boar taint. Six out of 8 slaughterhouses accepted intact or immunocastrated boars and the number of these animals per week ranged from 180 to 3,200 per slaughterhouse. Based on numbers of weekly slaughterings of all male pigs, 3.3% were intact and 4.9% immunocastrated boars. Main reason for the decision of slaughtering intact or immunocastrated boars is on demand of the customer. One of the 2 abattoirs slaughtering only castrated males does not consider to change its strategy in the future because of a lack of a potential market. The other will change its strategy in 2014. The results of the questionnaire indicated a clear need for a reliable boar taint detection system which should score on a 2-point scale; presence of absence of boar taint. It should be preferably located immediately after veterinary inspection at the slaughter line. Results should be reported to the pig producer. Other preferred characteristics of a future detection system should be: (1) easy adaptation to changing boar taint parameters; (2) easy to clean and safe; (3) no interference with food safety; (4) full automatic; (5) acceptable for customers buying the intact boar meat; (6) results available immediately after analysis. Only 2 slaughterhouses currently used a boar taint detection system, which twice was the sensory method. Both abattoirs analyse all intact males, one does the analysis on the carcass, the other in the laboratory. Slaughterhouses abilities to change over towards more weekly slaughtering of intact boars is mainly controlled by the consumers market. The absence of reliable boar taint detection systems at the slaughterhouse acts as a brake. Notwithstanding these limitations, Belgium has a unique position in Western-Europe as the only country marketing intact boars and immunocastrates.

Welfare of sows and neonatal piglets in SWAP pens

J. Hales[1], V.A. Moustsen[2], M.F. Nielsen[2] and C.F. Hansen[1]
[1]Department of Large Animal Sciences, Groennegaardsvej 2, 1870 Frederiksberg, Denmark, [2]Pig Research Centre, Axeltorv 3, 1609 Copenhagen, Denmark; hales@sund.ku.dk

Housing sows in farrowing crates can increase physiological stress and have detrimental effects on sow welfare. However, in farrowing pens piglets may have greater risk of dying, especially in the first days after farrowing. The objective of this study was to assess the welfare of sows and piglets in SWAP (Sow Welfare And Piglet protection) pens where sows could be confined. The study was conducted in a 1,200 sow piggery with Danish Landrace × Yorkshire sows. Sows were randomly allocated to one of three treatments. Loose-loose (LL) were loose from placement in the farrowing unit to weaning, loose-confined (LC) were loose from entry to birth of the last piglet (BLP) and confined from BLP to day 4 post farrowing, and confined-confined (CC) were confined from day 114 of gestation to day 4 post farrowing. All sows were loose housed after day 4. Live born and stillborn piglets were recorded for 365 LL-, 358 LC- and 342 CC-sows. All dead piglets were collected for post-mortem examination. Stress assessment using heart rate variability and saliva cortisol was performed on a subsample of sows (48 sows/group). A Polar heart rate monitor was fitted day 114 of gestation to day 4 post farrowing. Saliva was collected at 8 h, 13 h and 16 h from day 114 of gestation to day 4 post farrowing. All data were analysed using generalised linear models and results are presented as estimates (95% CI). Piglet mortality from litter equalization to day 4 was lower in CC (5.9% (5.2-6.7)) than in LC (6.9% (6.2-7.8); P<0.05) and in LL (8.0% (7.2-8.9); P<0.001). From day 4 to weaning, piglet mortality did not differ between treatments (5.7-6.0%; P=0.87). Cortisol level increased in CC from 19.6 nmol/l (15.8-24.2) day 0 to 27.2 nmol/l (22.0-33.6) day 1 and decreased thereafter to 24.1 nmol/l (19.4-29.8) day 2, and 20.4 nmol/l (16.5-25.2) day 3. In LC levels rose from 32.3 nmol/l (26.5-39.5) day 0 to 36.9 nmol/l (30.2-45.1) day 1 and decreased to 28.7 nmol/l (23.5-35.1) day 2, and 27.2 nmol/l (22.3-33.2) day 3. Levels in LL increased from 40.1 nmol/l (32.7-49.2) day 0 to 47.5 nmol/l (38.7-58.3) day 1 and then decreased to 41.1 nmol/l (33.4-50.6) day 2, and 33.3 nmol/l (27.1-41.0) day 3. Sows in LC had higher levels than CC on day 0 (P=0.001) and day 1 (P=0.04). Sows in LL had higher levels than LC on day 1 (P=0.02) and day 2 (P=0.02). Compared to CC, LL had higher levels on day 0 (P<0.001), day 1(P<0.001), day 2 (P<0.001), and day 3 (P=0.001). On day 4 levels were similar (21.6-28.4 nmol/l). In conclusion, there were differences in sow and piglet welfare between all systems. Sows in LL had greater piglet mortality and higher cortisol concentration the first 4 days post farrowing than sows in LC and CC.

Poster 148

Space allowance for cattle in feedlots and animal welfare: health indicators

F. Macitelli[1,2], J.S. Braga[1] and M.J.R. Paranhos Da Costa[1]
[1]*Universidade Estadual Paulista, Zootecnia, Rod Paulo Donato Castelane s/n, 14884-900 Jaboticabal, Brazil,* [2]*Universidade Federal de Mato Grosso, Zootecnia, rodovia mt 270, 78700-000 Rondonópolis, MT, Brazil; macitellif@gmail.com*

The aim with this study was to evaluate the effect of space allowance for cattle in feedlots on animal welfare using health indicators. Nellore and crossed bulls (A. Angus × Nellore), with 22.0±4.0 months old on average, were confined for 88 days. Three treatments were tested, varying the space allowance per animal (T1=6, T2=12 and T3=24 m^2/animal), and three lots (150 animals each) were observed for each treatment (n=1,350). The percentages of animals with nasal (ND) and ocular discharge (OD) and number of sneezing (SNE) and coughing (COU) per lot/minute were recorded in each feedlot pen weekly, during two periods, drought and rainy. The percentages of animals with macroscopic signs of pulmonary emphysema and bronchitis were recorded per treatment, just after slaughter. The variance analysis was used to analyses the effects of treatments and periods (as fixed effects), besides the interaction between them, on ND and OD (using Proc. GLM, SAS) and SNE and COU (using Proc. GENMOD SAS). Qui-square test was used to evaluate the frequency of bronchitis cases variation, applying SAS FREQ procedure. The interaction effect between treatments and periods did not affect any variable analysed. Treatment affected significantly ND (P=0,018), with means (±SD) of 33.3±21.5, 24.4±13.1, and 12.2±9.8%, and COU (P=0,052), with means (±SD) 1.3±0.5, 0.3±0.2 and 0.0±0.0 coughs/min.; for T1, T2 and T3, respectively. There was no significant variation between treatments in OD and SNE, although in the last case there was a relevant numeric variation between the treatments means (±SD) (1.7±2.1, 0.8±1.0 and 0.5±0.6, sneezes/min., for T1, T2 and T3, respectively). There was significant variation (P<0.05) between the periods for all variables, averaging 2.0±1.3 and 0.0±0.0 sneezes/min, 0.0±0.0 and 0.6±1.0 coughs/min, 32.6±19.0 and 14.1±8.5% of animals with ND and 26.7±5.6 and 10.5±8.2% of animals with OD, in drought and rainy periods, respectively. A significant (P<0.05) variation between treatments was found in the percentage of animals diagnosed with macroscopic cases of bronchitis (X^2=45.5), with 16.0, 9.0 and 1.0% of animals, for T1, T2 and T3, respectively. Macroscopic cases of pulmonary emphysema were identified only at T1 (3.0%). These results indicate that the space allowance in feedlots plays important role in the definition of the cattle welfare status, since low space availability increases the risk number of animals showing signs of health problems, which was confirmed by the number of cases of chronic obstructive pulmonary disease.

Estimating the economic value of Australian stock herding dogs

E.R. Arnott, J.B. Early, C.M. Wade and P.D. Mcgreevy
The University of Sydney, Faculty of Veterinary Science, The University of Sydney, NSW 2006, Australia; elizabeth.arnott@sydney.edu.au

To maximise profitability, producers must be cognisant of the cost of production and make investment decisions based on the expected financial returns. Approximately 270,000 herding dogs employed throughout Australia, represent an investment into farm labour efficiency. Therefore, expenditure decisions associated with the care and upkeep of working dogs should be informed by knowledge of the value of these animals. In this way, the welfare of the farm working dog is intimately linked with their perceived value. The goal of the present study was to estimate the net economic worth of the Australian stock herding dog. Additionally, we hoped to gain some insight into the way farmers currently perceive the worth of their stock dogs by assessing financial decisions that directly affect their dogs. Data on a total of 4,027 dogs were acquired through The Farm Dog Survey, a questionnaire which gathered information from 812 herding dog owners around Australia. To estimate the typical economic contribution of the dogs, the median values for the major costs associated with dog ownership were summated and compared to the median number of hours worked over a lifetime by the sample of dogs reported in the Farm Dog Survey. To create a financial representation of time investments and returns, an hourly rate of $20 (the typical farm hand wage) was used. Results: The median cost involved in owning a herding dog was estimated to be a total of AU$7,763 over the period of its working life. The work performed by the dog throughout this time was estimated to have a median value of $40,000. So, herding dogs typically provided their owners with a 5.2 fold return on investment. When respondents were asked to nominate the maximal veterinary expenditure they would consider for an especially valued dog, the median response was AU$1,001 – AU$2,000 which is not concordant with the dogs' calculated median lifetime value. Conclusion: While the costs associated with acquiring and keeping stock herding dogs are minimal, such modest costs should not be considered a reflection of their worth. Estimates of the financial contribution of the typical working dog to the farmer indicate at least a five-fold return on investment. The expenditure decisions of working dog owners in the survey do not reflect recognition of the value of these dogs. Therefore, these findings may serve to equip working dog owners with useful information to make financially appropriate expenditure decisions when it comes to their working dogs. This could lead to increased profitability for farmers and improved welfare for their dogs.

Effects of space allowance and a straw rack in commercial farms on tail lesions in fattening pigs

K. Schodl[1], C. Leeb[2] and C. Winckler[2]
[1]University of Natural Resources and Life Sciences (BOKU), Doctoral School of Sustainable Development (dokNE), Peter-Jordan-Str. 82, 1190 Vienna, Austria, [2]University of Natural Resources and Life Sciences (BOKU), Department of Sustainable Agricultural Systems, Gregor-Mendel-Str. 33, 1180 Vienna, Austria; katharina.schodl@boku.ac.at

Tail biting in fattening pigs presents a critical issue regarding animal welfare as well as economic and production aspects on pig farms. In conventional pig fattening tail docking has become a widely applied measure to prevent tail biting. However, cutting of the tail is a painful procedure for the piglets and impairs animal welfare. Despite the multifactorial nature of tail biting, several studies show that lack of adequate manipulable material and of space are highly important factors, which is often the case on commercial fattening farms. In the course of a larger study a list of measures with the aim to improve animal welfare on conventional pig fattening farms was implemented on three farms in Austria. Among others, the measures comprised an increase of space allowance per animal (1 m^2 instead of 0.7 m^2/pig) and the provision of straw in racks as manipulable material. Implementation took place in half of the pens (test pens) whereas the remaining pens served as a control. On Farm 1 (974 pigs, 74 pens) all pigs were tail docked, pigs on Farm 2 (413 pigs, 47 pens) were tail docked in the control pens but had intact tails in the test pens and on Farm 3 (70 pigs, 10 pens) all pigs had intact tails. The aim of this paper is to examine the effect of the measures on the occurrence of tail lesions and to find out whether tail length has an influence on the prevalence of tail lesions (Farm 1 and 3). For this purpose an assessment matrix was developed which allows to record tail length and degree of injury for each animal in a pen. Pens were assessed at the beginning and in the middle of the fattening period. Moreover, tails have also been scored at the slaughterhouse using the same scoring system. Data were analysed for each farm individually. Differences in tail lesions between test and control pens, between the first and second observation as well as differences in the prevalence of tail injuries for 5 categories of tail length were analysed using non-parametric tests. The mean prevalences of tail lesions at the beginning of the fattening period were 4.79, 5.21 and 3.67% for Farm 1, Farm 2 and Farm 3, respectively. Analyses of the on-farm data revealed no significant differences for all comparisons. Hence, it can be concluded that the increase of space allowance and provision of a straw rack has no influence on the prevalence of tail lesions, independent from tail length or whether tails are docked or not. Slaughterhouse data will subsequently be analysed.

Influence of age on the use of different environmental enrichments in early-weaned pigs

M. Battini[1], S. Barbieri[1], E. Neri[1], M. Borciani[2], A. Gastaldo[2] and E. Canali[1]
[1]Università degli Studi di Milano, DIVET, Via Celoria 10, 20133 Milan, Italy, [2]Centro Ricerche Produzioni Animali, Viale Timavo 43/2, 42121 Reggio Emilia, Italy; monica.battini@unimi.it

The intention of the European legislation on rooting materials is to improve the welfare of pigs, supporting their need to express exploratory behaviours (i.e. rooting, sniffing, and chewing), reducing the risk of pigs performing abnormal behavioural patterns that may impair welfare and production. This assumes remarkable importance in case of early-weaned piglets (3 weeks of age), where explorative and manipulative behaviours (especially suckling behaviours) are strongly needed. The aim of this project is to gather information about the effect of period provision of environmental enrichments on early-weaned piglets' behaviour. The study was conducted in a commercial farm in Northern Italy. Five groups were created (four enriched groups and one control group with no enrichment, 6-8 piglets each) at two age periods (P1: 3 weeks of age; P2: 5 weeks of age). Four different environmental enrichments (wood, chain, sisal rope, hemp rope) were presented to the enriched groups, both in P1 and P2. Behavioural observations were performed five days after the enrichments introduction, using instantaneous and scan sampling method (60 min/10 min scan) on every five groups. The main behaviours recorded were: use of enrichment (UE), manipulating the box (MB), manipulating other piglets (MP), and aggressive behaviour caused by the enrichment (AB). Behavioural data were analysed by GLM repeated measures to evaluate the effect of enrichment, period and their interactions. Although no significant differences have been found among enrichments, results show that weaners prefer the chain, whereas sisal and hemp ropes are scarcely used, both in P1 and P2. UE is significantly higher in P1 ($P<0.01$): the enrichments provided stimulate suckling behaviours, strongly needed at 3 weeks of age. MB is significantly higher ($P<0.001$) in P2 (except for hemp rope), due to the increase of exploratory behaviours (e.g. rooting for feeding) at 5 weeks of age. MP decreases from P1 to P2 (except for sisal rope), but no significant differences have been found. AB has been recorded in none enriched groups: this supports the idea that one point-source enrichment object can be adequate for 6-8 pigs. Irrespective of the type of enrichment, the study shows that manipulable materials are preferred by 3-week-old compared with 5-week-old pigs. Furthermore, the exploratory behaviours directed to other piglets confirm the strong need of manipulable objects on early-weaned piglets at 3 weeks of age. Further studies are required to gather information about appropriate environmental enrichment materials at different stages of age.

Impact of mixing prior transport to slaughter on skin lesions in boars

N. Van Staaveren[1,2], A. Hanlon[2], D. Teixeira[1] and L. Boyle[1]
[1]Teagasc, Pig Development Department, Animal and Grassland Research and Innovation Centre, Moorepark, Fermoy, Co. Cork, Ireland, [2]University College Dublin, School of Veterinary Medicine, Belfield, Dublin 4, Dublin, Ireland; nienke.vanstaaveren@teagasc.ie

Post-mixing aggression is associated with increased skin damage which is a welfare related problem but can also lead to carcass downgrading and financial losses for the producer. Meat inspection has been suggested as a tool to monitor herd health and welfare. Our aim was to investigate the effect of mixing prior transport to slaughter on skin lesions in boars. Three hundred pigs were slaughtered in 5 replicates. Pigs from three single sex groups were randomly allocated to one of three treatments (n=20) on the day of slaughter: male group unmixed prior transport (MUM), male group mixed prior transport (MM), and male and female group mixed prior transport (MF). Skin lesions were assessed on all experimental pigs on day –5 and day – 1 of slaughter, at lairage (6 focal pigs/treatment) and after slaughter on a 0-5 scale. Loin bruising was scored on all carcasses on a 0-2 scale. Data were analysed in SAS (total lesion scores: PROC MIXED; scores per body part: PROC GENMOD; associations between lesions: PROC CORR). No significant effect of treatment or sex was found on total skin lesions scored on farm and lairage. However, total skin lesion score was significantly higher at lairage than at day -1 (8.42 vs 6.01, s.e.m. 0.45; P<0.0001). Measurement and treatment interaction tended to significantly affect total lesion score (P=0.08), showing a smaller increase from farm to lairage for MUM pigs than MF and MM pigs and lesion score increasing more strongly for MM pigs compared to MF pigs. At the carcass no significant effect of treatment or sex was found on total skin lesion scores, separate body part scores and loin bruising (P>0.05). Total skin lesion scored on farm was positively correlated with skin lesion scored at lairage (r=0.45, P<0.0001) and on carcass (r=0.21, P<0.05). These results suggest that mixing pigs increases skin damage and mixing boars with other boars is more detrimental than mixing boars with gilts. At carcass these differences are less obvious probably due to changes in appearance of lesions as the carcass is handled and cleaned down the slaughter line. Carcass skin lesions are correlated with farm lesions suggesting that meat inspection can be used to inform farmers about herd health and welfare.

Assessment of piglet welfare following castration with and without analgesia

P.V. Turner[1], A. Viscardi[1], P. Lawlis[2] and M. Hunniford[3]
[1]University of Guelph, Dept. of Pathobiology, OVC, Guelph, ON N1G 2W1, Canada;
[2]Ontario Ministry of Food and Agriculture, 1 Stone Rd W, Guelph, ON N1G 4Y2, Canada;
[3]University of Guelph, Dept. of Animal and Poultry Sciences, Guelph, ON N1G 2W1, Canada;
pvturner@uoguelph.ca

There is a critical lack of information surrounding methods to improve the wellbeing of piglets undergoing painful procedures, such as castration, which occur as part of routine on-farm management practices. This has made it challenging to promote alternative methods that reduce pain and distress in swine. One method used commonly to eliminate surgical pain in companion animal species and humans is to administer a topical anaesthetic agent (EMLA; prilocaine-lidocaine cream) together with a long-acting nonsteroidal analgesic agent, such as meloxicam (MEL). We evaluated the individual and combined effects of EMLA and meloxicam, in alleviating pain and distress associated with castration in piglets as measured by assessing general behaviour (time spent performing a range of 17 discrete behaviours including standing, lying, suckling and vocalizations). Four litters of piglets were used (~5 piglets/litter were male, n=19) and treatments were randomized across litters. Treatment groups consisting of MEL 0.4 mg/kg IM, EMLA topical, MEL IM + EMLA topical, or nothing administered 15 minutes prior to castration. Castration was performed by skilled personal using an open castration technique and a scalpel to transect the spermatic cord and vessels. Animals were videotaped at -24 h, 0, 1, 2, 3, 4, 5, 6, 7, and 24 h after castration. Behaviours of each piglet were scored continuously from videotapes for the first 10 minutes of each hour by an observer without knowledge of animal treatment. Behaviours were evaluated for normality and a mixed, 2-way ANOVA model was used for analysis. Except for tail wagging, which was found to occur more often in piglets receiving EMLA or EMLA + MEL ($P<0.02$), there was no significant beneficial effect of treatment on behaviour of piglets post-castration. All castrated piglets displayed significantly decreased activity, increased lying and sleeping, decreased nosing and exploration, increased stiffness, and decreased standing ($P<0.0001$) following castration and these effects persisted up to 7 h. These results indicate that castration is painful in piglets and the effects last for many hours post-procedure, with a normal repertoire of behaviours returning approximately 7 h after castration. Pain is not mitigated by the use of a long-acting nonsteroidal anti-inflammatory agent, meloxicam, alone or in combination with a topical anaesthetic agent (EMLA). Current recommendations for treatment of piglets undergoing castration, including dose of meloxicam, may require review in light of these findings.

Genetic parameters for arena behaviour of 8-month-old Merino lambs

J.J.E. Cloete[1], S.W.P. Cloete[2,3] and A.C.M. Kruger[2]
[1]*Elsenburg Agricultural Training Institute, Private Bag X1, Elsenburg 7607, South Africa,*
[2]*Directorate: Animal Sciences: Elsenburg, Private Bag X1, Elsenburg 7607, South Africa,*
[3]*Department of Animal Sciences, University of Stellenbosch, Private Bag X1, Matieland 7602,*
South Africa; jasperc@elsenburg.com

Australian research indicates that temperament is related to the ability of ewes to rear their lambs in extensive sheep. Divergent selection on temperament led to an unresponsive genotype as compared to an exceedingly nervous genotype. Lambs from the unresponsive line survived better than those from the excitable line. This paper reports the responses of 8-month-old Merino lambs to a human being in an arena test. These lambs were the progeny of lines that were divergently selected for the ability of ewes to rear multiple offspring. A total of 2,617 animals born from 2001 to 2013 were assessed in a modified arena test, involving the placing of an individual animal in an arena of 10.6×4.0 m. The floor of the arena was marked out in 18 equal numbered squares. At one end of and outside the arena was a pen containing six to seven contemporaries of the test animals. An operator sat on a chair in the arena directly in front of this pen. A second operator introduced the test sheep to the arena at a distance of 10.6 m from the human seated inside the arena. The test sheep remained in the arena for three minutes and was observed by at least two recorders located in a building overlooking the arena. The presence of the animal in a specific square was recorded every 15 seconds. The following parameters described the behaviour of the sheep: the mean distance from the person, being averaged across the twelve recordings and the total number of boundaries crossed between squares as an indication of the total distance travelled. Other data recorded were the number of bleats and the number of times an animal urinated or defecated. Heritability estimates were 0.13 for the average distance from the human, 0.24 for the number of crosses, 0.34 for the number of bleats uttered, 0.13 for the number of times urinated and 0.03 for the number of times defecated. Genetic correlations indicated that animals that kept a greater distance between themselves and the operator crossed more lines (0.31) and defecated more often (0.69). Genetic trends indicated that animals in the line selected against their ability to rear multiples increased their distance from the human operator with time, while the positively selected line maintained the same distance throughout. The numbers of times urinated and defecated showed divergence similar to that observed for lamb rearing ability with time. Lower levels of stress during unfamiliar procedures by animals selected for an improved reproduction rate are important from an animal welfare perspective.

Influence of radiofrequency electromagnetic fields on chicken embryo

K. Pawlak and B. Tombarkiewicz
University of Agriculture in Krakow, Department of Animal Hygiene and Breeding Environment, Al. Mickiewicza 21, 31-120 Kraków, Poland; rzpawlak@cyfronet.pl

During the course of evolution, living organisms have developed in the constant presence of natural electromagnetic fields (EMF). Today, artificial electromagnetic fields which are a consequence of human progress begin to play a considerable role in shaping the Earth's electromagnetic environment. Electromagnetic smog can influence on health and welfare of animal. Due to the rate and specific characteristics of development and the well understood process of embryogenesis, chick embryo is frequently used as a model in different kinds of biological research, including studies investigating the effect of EMF on living organisms. Therefore, this study attempted to determine the effects of the 1,800 MHz field on chicken embryogenesis. Hatching eggs of Ross 308 line (n=100) were used in the experiment. The eggs were randomly divided into two equal groups and incubated under standard conditions in a laboratory incubator. Group I (control group) was incubated in control conditions, i.e. in the incubator without an EM field generator. Group II (1,800 MHz group)- chicken embryos were exposed to 1,800 MHz electromagnetic fields with power density of 0.1 W/m^2,10 times/day for 4 min throughout whole period of incubation. The body weight, heart weight, thickness of left and right heart ventricles, thyroid hormone levels (thyroxine; T_4 and triiodothyronine; T_3) were investigated in embryos in 12 and 18 day of incubation, newly hatched chicks and birds that are ready for slaughter. Moreover, the hatchability was also determined after the incubation process. The results showed that T_4 and T_3 concentrations decreased markedly in embryos and in newly hatched chicks exposed to EMF during embryogenesis. Differences in thyroxine, triiodothyronine concentrations between the EMF-exposed group and the embryos incubated without additional EMF were highest in newly hatched chicks. The additional EMF had no significant effect on the body weight of embryos and chicks. In 12, 18 days-old-chicken embryos and new hatched chicks exposed to EMF we found increase in thickness left ventricle but decrease relative heart weight. The results showed that the EMF significantly accelerated the time of hatching. In II group (1,800 MHz) it was 24 hours earlier in comparison to the control group. However, percent of hatched chicks in experimental groups with the additional EMF (86.9% for Group II) was close to the control group (87.2%). This study was performed under the project NN311536340 'Chicken embryo as a model in studies on the influence of radio frequency electromagnetic fields on the embryogenesis process'

The effect of environmental enrichment on the behaviour of beef calves

A. Bulens[1,2], S. Van Beirendonck[2], J. Van Thielen[2] and B. Driessen[2]
[1]*KU Leuven, Research Group of Livestock Genetics, Department of Biosystems, Kasteelpark Arenberg 30, 3001 Heverlee, Belgium,* [2]*KU Leuven|Thomas More Kempen, Group Animal Welfare, Kleinhoefstraat 4, 2440 Geel, Belgium; anneleen.bulens@thomasmore.be*

The objective of the present study was to investigate possible effects of environmental enrichment on the behaviour and extent of coat contamination of beef calves. In total, 180 Holstein beef calves were followed for a period of 20 weeks from the age of two weeks. Feeding ration and housing conditions were similar for all calves. After arriving on the farm, calves were given an adaptation period of four weeks. After this period, they were divided into three groups: pen with Jolly Ball™ (10 pens), pen with cattle brush (10 pens) and barren pen (control group) (10 pens). Each pen housed 6 calves. Jolly Balls™ were hung at an altitude of 1.3 m, in the middle of the pen. Cattle brushes were attached to the pen wall at an altitude of 1.2 m. Behavioural observations were carried out two days a week, according to a scan sample method. All observations were performed by the same observer and the ethogram used was based on existing literature. Next to behaviour, the extent of coat contamination of the calves was scored three times throughout the study. Data were analysed using the statistical software program SAS 9.3. Behavioural data lacked normality and were therefore analysed using a logistic mixed model. Contamination data were analysed using a Fisher's Exact Test. Behavioural results indicate that calves of the control group, without environmental enrichment, lied down more than calves in the enriched pens ($P<0.0001$). They also ran ($P<0.0001$) and jumped around ($P<0.0001$) less than the other calves. Results also revealed a difference in social behaviour; animals in enriched pens displayed more social behaviour ($P<0.0001$) than calves in non-enriched pens. There were also differences between the different forms of enrichment. In pens with a cattle brush, animals self-groomed less ($P=0.0004$) than those in control or Jolly Ball™-pens. However, there was no difference in playing behaviour between calves with the cattle brush or the Jolly Ball™. Results also indicated no association between treatment groups concerning extent of coat contamination. As a conclusion, environmental enrichment in the form of Jolly Balls™ or cattle brushes did not influence the extent of coat contamination of beef calves but did have a positive influence on their behaviour. Calves in enriched pens displayed more playing and social behaviour compared to calves in non-enriched pens. The two forms of enrichment tested in this study both equally induced play behaviour in calves, so both forms can be applied successfully to enrich the otherwise barren environment of beef calves.

Interactions of growing pigs with four different applications providing straw

A. Bulens[1,2], S. Van Beirendonck[2], J. Van Thielen[2], N. Buys[1] and B. Driessen[2]
[1]KU Leuven, Research Group of Livestock Genetics, Department of Biosystems, Kasteelpark Arenberg 30, 3001 Heverlee, Belgium, [2]KU Leuven|Thomas More Kempen, Group Animal Welfare, Kleinhoefstraat 4, 2440 Geel, Belgium; anneleen.bulens@thomasmore.be

Pigs are often housed in barren intensive housing systems but according to council directive 2008/120/EC, they must have access to enrichment materials. In order to be successful, enrichment should be destructible, edible, nutritional and dung-free. Straw seems to be an interesting enrichment material but is not a practical option for fully-slatted systems. Producers must therefore provide straw applications to reduce straw waste through the slatted floor. Both the type of straw used and the design of the application might influence the interest for and the effectiveness of the materials. It is therefore designated to verify differences in ways to present enrichment materials, in order to detect applications which are less interesting for pigs. To this end, four straw applications were tested simultaneously (1 enrichment application per pen) during 4 cohorts, presenting them each time during two weeks to 6 pigs (total n=96) in pens with slatted floors. The four applications tested were a straw rack (long-stemmed straw), a straw feeder (long-stemmed straw), a MIK Toy (in rolls pressed chopped straw) and the Ikadan Straw Dispenser with chopped straw. All pigs had ad libitum access to straw from one of the applications. Behaviour was recorded using video cameras. Data were analysed per individual pig for week 1 and week 2 separately using the logistic mixed model. The daily used straw quantity per pen was registered during the entire test period. The quantity of straw use was the highest in presence of the straw rack (P<0.0001), with an average straw use of approximately 2 kg/pen (6 pigs) per week. Straw use did not differ between week 1 and week 2 for any of the applications, but manipulative behaviour towards the application decreased for all applications during week 2 compared to week 1. This suggests a decrease in the required time to intake straw from the applications. In presence of the straw feeder, most of the time only one pig was occupied with the application, while in presence of the straw rack, the MIK Toy and the Funbar, more pigs interacted simultaneously with the device (P<0.0001). Pigs spent more consecutive scans interacting with the straw rack and the straw feeder (P=0.0003), which could be explained by the use of long straw with these applications. The results show that the duration of interactions with the applications might depend on the type of straw that is used. This study was funded by the Federal Public Service of Health, Food Chain Safety and Environment (Contract RT 12/02 SUCANNIB).

Effect of long-term weakened geomagnetic field on selected breeding parameters

B. Tombarkiewicz and K. Pawlak
University of Agriculture in Krakow, Department of Poultry, Fur Animal Breeding and Animal Hygiene, a. Mickiewicza 21, 31-120 Krakow, Poland; rztombar@cyfronet.pl

Geomagnetic field (GMF) is one of omnipresent factors of the environment, and is essential for normal development of living creatures. A substantial part of human population is exposed to risk of continual disarrays in GMF generated by ubiquitous steel elements in almost every building and means of transport. These problems concern also the livestock animals especially those kept in the factory farming conditions. The aim of this study was to examine the influence of long-term and prolonged for several generations deprivation of GMF on the selected breeding parameters of laboratory rats. Three generations of Wistar laboratory rats (F1, F2 i F3) were used in the experiment. The animals from experimental groups were placed in hypogeomagnetic conditions (component vertical value of GMF below 20 nT) whereas control groups were kept in conditions free of field disturbances – component vertical value of GMF about 38,000 nT [Geo-Scanner BPT 3010]. To estimate breeding parameters the following indicators as: number of animals born and reared in a litter, litter weight, estimated birth weight and weight gains in the 1st, 5th, 10th, 15th and 21st day of life were taken into account. The number of animals in a litter in the first two generations (F1 and F2) was noticeably lower but statistically not significant in the experimental groups compared with the control one. The average litter weight was also lower in the experimental groups and was statistically significance in F2 generation. Great differences in the numbers of born and reared animals were also observed in the litters of F1 and F2 generations. A great mortality of newborns from the experimental groups should be emphasized. In F2 generation it proved to be statistically significant. Rats mortality in the experimental group of F1 generation was 8.6% (1.8% in control group), F2 generation – 14.6% (2.3% in control group) and F3 generation – 10.3% (4.4% in control group). Average fertility and litter weight were higher in the experimental group of F3 generation. The average birth weight in the first two generations was similar in both groups, whereas in the experimental group of F3 generation it was significantly lower than in the control one. In the hypogeomagnetic conditions body weights in the first days of life of newborns from F1 and F3 generations were slightly lower (statistically insignificant) compared with those from the normal conditions. In F1 generation, from the fifth day of life body weights of animals from the experimental group were equal to the control one whereas in F3 generation they remain lower until the end of the experiment, being statistically significant in many cases. This study was supported by DS 32-10/KHDZFiZ

Cortisol response to stress and behavioural profile in dairy ewes

M. Caroprese, M. Albenzio, M.G. Ciliberti, R. Marino, A. Santillo and A. Sevi
University of Foggia, SAFE, Via Napoli, 25, 71122 Foggia, Italy; mariangela.caroprese@unifg.it

The first response to stress is the behavioural one which can be considered the cheapest response in terms of biological cost for the animals. The existence of a relationship between cortisol levels, after an acute stress, and behavioural profile in sheep was studied. An initial flock of 30 Comisana ewes was involved the experiment, and each of the 30 ewes was individually subjected to an isolation test in a novel environment. Subsequently, from the initial flock, two groups of 8 Comisana ewes each were retrospectively selected, and the animals were divided, according to their cortisol concentration 10 min after the isolation test, into HC (High Cortisol) ewes, having a peak of cortisol concentration >90 ng/ml (average: 119.3±11.8 ng/ml), and LC (Low Cortisol) ewes having a peak of cortisol concentration <80 ng/ml (average: 52.4±11.8). During the isolation test the behaviour of each animal was video recorded and behavioural activities, such as the latency time to movement, the duration of movement, and of the exploratory activities were registered. In addition, the number of bleats and of the attempts to climb over the pen fence, the number of defecations and urinations for the 10 min-period were recorded. During the isolation test, HC ewes exhibited shorter duration of movement and lower number of bleats than LC ewes. Results suggest that the extent of hypothalamic-pituitary-adrenal axis activation is responsible for alterations of behavioural responses. A hyperactive hypothalamic-pituitary-adrenal axis may contribute to express a passive coping style during stressful situations.

Poster 160

Welfare strategies to prevent and control FMD in bovines

A. Natarajan, V. Ganesan and N. Bharathi
TANUVAS, VUTRC, 4/221, Panduthakaranpudur, Karur, 639006 TamilNadu, India;
akila@tanuvas.org.in

Foot-and-mouth disease, the most devastating and highly contagious viral disease of cloven-footed animals, causes suffering to animals, death of young ones, loss of milk, meat and drastic fall in productive performances. The current outbreak of Foot and Mouth Disease (FMD) in 2013 has reached epidemic proportions in Tamil Nadu, Southern India caused death of about more than 12000 bovines including buffalos. It has significantly hit milk production, prompting dairy farmers to seek intervention, including vaccination on a war-footing. The sufferings and death of cattle and buffalo due to FMD continues even after the efforts of vaccination campaign by the Animal Husbandry department. A case study done in Karur district of Tamilnadu clearly stated that the outbreak of disease was due to arrival of diseased animal to livestock markets from the neighbouring state and virus transmission through river beds. The animals affected before the vaccination campaign started by the veterinarians. The outbreak was severe in the places where the animals tied in a confined area. The lack of awareness with the farmers in vaccination, bio security measures, ethno veterinary practices made easier for the spread of disease on a faster manner. Lack of sufficient veterinarians to treat the affected animals and lack of enough space to isolate the animals were also the supporting reasons. All this led to suffering to animals and economic loss to the owners. Though vaccination carried out every year, FMD outbreak occurred due to one or other reason. To prevent this disease vaccination drive should be conducted for cattle on a particular day in all parts of the country. Educating farmers towards the importance of vaccination, shed cleanliness and management care of affected animals is equally plays a major role in controlling the disease. Bio Security measures killed the virus and preventing the spread to other animals. Also disease screening mechanisms is to be implemented in livestock markets. These measures may be helped in preventing the occurrence of the dreadful disease and save the animals.

Moving dairy cattle from tie-stalls to loose-housing: effects on animal welfare and production

G.H.M. Jørgensen, L. Aanensen and V. Lind

Bioforsk Nord Tjøtta, P.O. Box 34, 8860 Tjøtta, Norway; grete.jorgensen@bioforsk.no

The aim of this study was to investigate how the transition from tie-stalls to an automated loose housing barn would affect dairy cattle welfare. We studied health, production, cleanliness and behaviour in tie-stalls, during the transition period and 22 weeks after moving from one management system to another. We used video surveillance to record animal movement in the barn, scoring number of animals using the mechanical brush, concentrate feeders, automated milking system (AMS) and eating roughage. Instantaneous sampling every ten minutes were performed from video recordings and direct measures on animal health, cleanliness and hoofs were performed on each individual in tie-stalls, after eight and 22 weeks in the loose housing system. Activity loggers (IceTag) on focal animals recorded cow activity. Differences in milk production of same individuals in the two housing systems were analysed using a paired T-test in Mintab 16. A total of 52 dairy cows from two different herds were introduced. During the first 24 h, we observed a large number of social interactions (running, mounting, head-butting, chasing) and skin scrapes and some individuals became lame. The cows increased their activity and spent less time lying (<8 h.). Within the first week, the average lying time increased from 8.5 to 12 h/day, but some individuals were lying in the dirty traffic alleys. Within a month, most of the dairy cows used the resting cubicles regularly and the average lying time increased to 14.5 h/day. Cows became dirtier on hindquarters and legs and less dirty on sides and bellies after moving to loose housing. Increased cleaning intervals on the floor scrapers and increased use of resting cubicles reduced the number of cows with high dirt scores within the 22 weeks. Cows in their first lactation interval (0 to 5 months after calving) decreased their production in loose-housing (mean ± SEM: 20.5±1.4 kg/day) compared to the tie-stalls (27.6±1.2 kg/day; T-value=4.9; P<0.0001). No such difference was found for cattle in late lactation interval (5 to 10 months after calving). Some cows refused to use the AMS voluntarily, became trapped in the waiting area without access to food or water for several hours each day. Such cows were culled and replaced independent of individual production level. We identified several bottlenecks that might affect health, production and animal welfare. We conclude that moving dairy cows directly from tie-stalls to loose housing combined with herd mixing is challenging for the individual cow. Farmers need to closely monitor all individuals and their ability to adjust to the technology during the first four weeks after introduction.

Farmer attitude affects the overall welfare of dairy cows

S.N. Andreasen[1], P. Sandøe[1], S. Waiblinger[2] and B. Forkman[1]
[1]University of Copenhagen, Department of Large Animal Sciences, Groennegaardsvej 8, 1870 Frederiksberg C, Denmark, [2]University of Veterinary Medicine Vienna, Institute of Animal Husbandry and Animal Welfare, Department for Farm Animals and Veterinary Public Health, Veterinärplatz 1, 1210 Vienna, Austria; sinen@sund.ku.dk

Previous studies have found that the attitude and behaviour of the farmer affects the animals. Farmers with a negative attitude toward animals display a more negative behaviour which e.g. induces stress as well as reduced productivity in the animals. The current study investigated whether farmer attitude affects the overall welfare of dairy cows in their care. Thirty-five Danish dairy farms were included in the study. The welfare of the cows was assessed using the Welfare Quality® (WQ) protocol. The attitude of the farmer was assessed using an attitude questionnaire previously used and validated by S. Waiblinger and co-workers. Using Principal Component Analysis 15 components were found to describe the attitude of the farmers in relation to their attitude toward the cows, their attitudes towards interacting with cows when moving them to and from milking and when milking, their attitudes toward contact with cows in general and their attitudes toward working with cows. The attitudes were correlated with the 12 criteria, the four principles and the overall score of the WQ and to the milk yield. A significant negative correlation was found between the overall welfare score and a farmer attitude where the farmers were in agreement with the statements of kicking or hitting (with a stick) cows if necessary to move them (r_s=-0.42, P=0.01). A significant moderate positive correlation was found between positive farmer attitude and WQ principle 4 (Appropriate behaviour), in which avoidance distance is included (r_s=0.35, P=0.004). In regard to milk yield the results showed that farmers agreeing more with negative handling of cows when moving them to and from milking correlated negatively with the milk yield (r_s=-0.47, P=0.03). Farms reaching a high score in WQ criterion 6 (Absence of injuries) and principle 3 (Good health) were the farms with the highest yield (r_s=0.36, P=0.03; r_s=0.36, P=0.03). These results indicate that negative farmer attitudes lead to poor overall welfare and reduced milk yield. A positive farmer attitude is related to more confident animals. Finally, having animals in good health increases the milk yield. These findings are in line with results from previous studies. These findings together with previous findings indicate that farmer attitude is important in animal welfare and for this reason knowledge of farmer attitude has a prospect for being included in educational programs for farmers and in interventions on-farm raising positive farmer attitude and animal welfare.

Spinal deformation, bone quality and pododermatitis in rabbit does: effect of housing and floor type

S. Buijs[1], K. Hermans[2], L. Maertens[1], A. Van Caelenberg[2] and F.A.M. Tuyttens[1]
[1]Institute for Agricultural and Fisheries Research, Animal Sciences Unit, Scheldeweg 69, 9090 Melle, Belgium, [2]Ghent University, Veterinary Faculty, Salisburylaan 133, 9820 Merelbeke, Belgium; stephanie.buijs@ilvo.vlaanderen.be

Reproduction does (the maternal parent stock of meat rabbits) are commonly housed in individual wire floor cages. Such housing is under increased scrutiny because it is thought to restrict activity and diminish foot health. We compared individual wire cage housing (INDIWIRE) to two alternative systems: semi-group housing on a wire floor (GROUPWIRE), and semi-group housing on a plastic slatted floor (GROUPPLAST). In the roofless, 2 m^2 semi-group housing pens, 4 does were housed communally during half of each 42-day long reproduction cycle (to increase the opportunity and incentive for activity). The pens were separated into 4 individual units during the other half of each cycle (around kindling, when aggression peaks). We used 6 GROUPWIRE and 6 GROUPPLAST pens. The 24 individual cages had a floor area of 0.4 m^2 and were 63 cm high at the highest point. All systems were equipped with a platform. Hycole does (29 weeks old) were allotted randomly to the systems 3 days before their 2nd kindling, and stayed within treatment until after the weaning of the 5th litter when we collected our data. We expected semi-group housing to decrease the prevalence of spinal deformations (scoliosis, kyphosis and lordosis, assessed by post-mortem X-rays) due to increased activity. However, we found no difference between the systems (binomial: $F_{2,62}=0.4$, P=0.68) although overall prevalence was high (38%). In contrast, the greater tibia cortex thickness in the semi-group pens did suggest that activity was increased in these systems (ANOVA: $F_{2,62}=3$, P=0.05, LSMEANS ± SEM: 1.45, 1.46 and 1.38±0.03 mm, for GROUPWIRE, GROUPPLAST, and INDIWIRE, respectively). Plastic flooring was expected to decrease pododermatitis, but true ulcerative pododermatitis was absent in our study (likely due to relatively young age of the does and the application of footrests to the wire floors). However, the prevalence of plantar hyperkeratosis (a very early stage of pododermatitis characterized by hair loss and callus formation) was much lower on the plastic floor (GROUPPLAST: 5±8%) than on the wire floor (65 and 68% ±9 for GROUPWIRE and INDIWIRE, respectively, binomial: $F_{2,62}=12$, P<0.0001). In conclusion, both semi-group housing and plastic flooring had a positive effect on one of our indicators. The exact impact on welfare requires further study because little is known about the progression of plantar hyperkeratosis into ulcerative pododermatitis and because the underlying reason for the suggested increased activity is unknown (i.e. it may be due to fleeing from aggressive pen mates).

Nociception and pain perception in noble crayfish (*Astacus astacus*) using the novel object paradigm

W. Hendrycks[1], T. Abeel[1], X. Vermeersch[2], E. Roelant[1] and H. Vervaecke[1]
[1]University College KAHO, Agro- and Biotechnology, Ethology and Animal Welfare, Hospitaalstraat 23, 9100 Sint-Niklaas, Belgium, [2]Institute for Agricultural and Fisheries Research, Technology and Food Science Unit, Brusselsesteenweg 370, 9090 Melle, Belgium; wouter.hendrycks@gmail.com

The perception of pain in animals, vertebrates and invertebrates alike, is triggered by specific neural pathways as a response to noxious stimuli (i.e. nociception). Little is known on the more conscious experience of pain perception in crayfish. In the paradigm of a novel-object-test, a decreased reaction towards a novel object is expected in animals that experience a temporary reduced environmental alertness due to pain. Animals in pain are expected not to react differently to a novel versus familiar object. We assessed the overall effect of electrical shocks on the noble crayfish (*Astacus astacus*) and determined how this was affecting the ability of the crayfish to focus their attention towards a newly introduced object. We also examined the difference in response towards a familiar object and an unfamiliar object. There were significant differences in activity among shocked and not-shocked crayfish. Noble crayfish who did not receive a shock did not approach the unfamiliar object in the first quarter after the introduction. Noble crayfish that were shocked approached objects significantly more and the latency time to approach this object was lower than in not-shocked crayfish, indicative of a lower degree of neophobia to a novel object after experiencing noxious stimulation. Crayfish that were outside their shelter after receiving a shock also appeared disorientated as they were sometimes seen to bump their head against the novel object or unable to enter their shelter. Nociception may capture the animal's attention with only little attention directed at responding to the fear of the novel object. Shocked crayfish showed more waving in the first minutes after the shock, a behaviour whose function merits further exploration. The results show the occurrence of nociception in crayfish and some evidence of a higher level experience of pain in crustaceans, evidenced by their altered state of alertness.

Poster 165

Do disbudded calves benefit from five-day pain medication?

J. Hietaoja, S. Taponen, M. Pastell, M. Norring, T. Kauppinen, A. Valros and L. Hänninen
Research centre for animal welfare, P.O. Box 57, 00014 University of Helsinki, Finland;
laura.hanninen@helsinki.fi

Disbudding causes pain, but the needed length of post-operative pain medication or the possible differences between disbudding with heat cauterization vs caustic paste are not known. We studied the differences between these methods and the effect of prolonged treatment with a non-steroidal anti-inflammatory drug (NSAID) on dairy calves´ activity and milk-feeder visits. We performed a 2×2 factorial design study on two-week-old dairy calves weighing (mean ± SE) 51±1 kg. We disbudded calves during sedation with either heat cauterization (n=24) or caustic paste (n=23). Cauterized calves were also given local anaesthesia. All calves got NSAID orally (4 mg/kg) before sedation, and the treatment group during four more days (n=24). Control calves the got equivalent volume of water (n=23). All calves were sham-disbudded 2 days prior (day −2) disbudding (day 0). We registered calves activity and milk-feeder visits with accelerometer and automatic-milk-feeder from day -2 to day 8. Calves were weighed days -2, 0, 7 and 10, and calculated number of nutritive and non-nutritive visits, consumed milk, and number and duration of lying bouts and total durations for daily lying and activity. The differences between treatments were analysed with repeated sampling linear mixed models. Disbudding method did not affect lying behaviours but NSAID tended to decrease daily overall activity (P<0.06): calves receiving NSAID for 5 days spent less time active than calves medicated only for 1 day (473±24 vs 530±22). Treatment×day interactions were found for number of non-nutritive and all feeder visits: calves medicated for 5 days had more visits on days 3 and 4 and non-nutritive visits on day 3 than calves medicated for 1 day (P<0.05 for all). Cauterized calves drank quicker than caustic paste disbudded ones (0.7±0.03 l/min vs 0.6±0.03 l/min; P=0.015). Cauterized calves grew overall better than paste disbudded calves (0.93±0.05 kg/d vs 0.88±0.05 kg/d) and calves medicated for 5 days grew better than calves medicated for 1 day (1.0±0.05 kg/d vs 0.6±0.05 kg/d; P<0.05 for both). Also a medication×method interaction was found (P<0.05): cauterized calves that were medicated for 5 days grew better than cauterized calves medicated for 1 day (1.1±0.1 vs 0.8±0.1 kg/d), no difference was found within caustic paste disbudded animals (0.9±0.1 vs 0.9±0.1 kg/d, respectively). Amount of consumed milk did not differ between treatments. Disbudded calves might benefit from a five-day NSAID medication, as it increased growth, reduced overall activity and increased number of feeder visits. Caustic paste calves had decreased growth and drinking speed, with no effect of NSAID.

Improvement in feather loss of laying hens is associated with a large scale welfare initiative

S. Mullan[1], D.C.J. Main[1], M. Fernyhough[2], S. Butcher[2], J. Jamieson[3] and C. Atkinson[3]
[1]University of Bristol Veterinary School, Somerset, BS40 5DU, United Kingdom, [2]RSPCA, West Sussex, RH13 9RS, United Kingdom, [3]Soil Association, Bristol, BS1 3NX, United Kingdom; siobhan.mullan@bristol.ac.uk

The inclusion of welfare outcome assessments within farm assurance schemes has been advocated to provide more robust welfare assurance than the traditional resource-based assessments. The Soil Association (SA) and RSPCA Freedom Food (FF) Schemes, in partnership with the University of Bristol, implemented welfare outcome assessments into their annual scheme visit on their laying hen farms with an additional aim of improving the welfare of the birds on their farms. By September 2011 all inspectors on FF and SA farms recorded observations of up to 50 birds, randomly selected, for signs of feather loss (slight or moderate/severe(m/s)) in two body regions: head and neck, and back and vent, and for signs of dirtiness. Other semi-quantitative and records-based measures were also included (see www.assurewel.org). The 50 bird sample was feasible to include within the existing audit time and provided a scheme prevalence accurate to within 0.9% (95% CI). The schemes recognised that improving welfare would require more than just observations so the results of the 50 bird sample, although not always able to provide an accurate farm prevalence, were used to stimulate benchmarking discussions with farmers to encourage changes to improve welfare and to identify farms that may benefit from further support. After 2 years of implementation data were available from 830 farms assessed in year 1 and 743 farms in year 2. Of these, 90% of farms were FF, 10% were SA, 81% of flocks were free-range, 17% were organic and 3% were barn, and 79% of flocks were beak trimmed. The mean age of the birds at assessment was 45 weeks and the mean flock size was around 7,750 birds. The number of birds recorded with feather loss reduced by a third from year 1 to year 2, from 32% (10% m/s) to 22% (6% m/s) for the head and neck region and 33% (13% m/s) to 23% (6% m/s) for the back and vent region. Zero-inflated models included the year of assessment as a significant explanatory variable for these improvements. The proportion of birds that were dirty remained small from year 1 (3%) to year 2 (2%). Changes aiming to improve welfare were made by 52% of farmers during this time as described by Butcher et al. This is the first time that large scale implementation of welfare outcome assessment and support has demonstrated an improvement in welfare, although it is recognised that during this period of time there were other industry initiatives aimed at reducing feather loss in addition to the awareness raising and targeted support available through the SA and FF schemes.

Frequent handling of cows at milking improve human-animal relationship in dairy cattle herds

N. Hald, T. Rousing and J.T. Sørensen
Aarhus University, Animal Science, Blichers Allé 20, P.O. Box 50, 8830 Tjele, Denmark; ninah.andersen@agrsci.dk

The human–animal relationship is an important factor when considering animal welfare in dairy cattle herds. Fearful animals being unpleasantly affected by recurring contact with humans explain the welfare relevance of a strained human–animal relationship. It is expected that this could be self-perpetuating. Furthermore, several studies indicate that a strained human–animal relationship may also depress the milk yield in dairy cows. This study investigated the possible relation between herd ratio of dairy cows avoiding humans in a human approach test and quality and quantity of tactile handling during milking. 18 Danish dairy herds – all exceeding 100 year cows – with a loose housing, primarily cubicle system (one herd had a deep litter system, another deep litter for some cow groups) and a traditional milking parlour where included in the study. The study was conducted during 2012 with a one-day visit in each herd. Cows avoidance distance (AD) where recorded in a human approach test performed in the cubicle loose housing system. Tactile contact between milker and the individual cows where quantified and qualified during morning milking as either 'neutral/positive' (gentle patting, resting hand), 'mildly negative'(moderate patting or push, gentle slap with hand or plastic stick or alike) or 'moderately negative' (hard patting or push, hard slap with hand or plastic stick or alike). Results showed that herd ratio of cows showing an avoidance distance less than or equal to 0.2 m to an approaching human (categorised as 'not fearful') varied between 16 and 91%, and further that there where a significant and positive effect on this measurement of quantity tactile human contact during milking ($R^2=0.78$). When number of interactions per cow per milking increased with 0,5 the ratio of cows not avoiding human in a human approach test (AD≤0.2 m) increased by 45 percentage points. However, the quality of the handling did not affect the human –animal relationship. These finding makes one speculate that if not severely negatively handled during milking (which none of the cows presumably were in the present study) – quantity of contact has an positive effect on the human-animal-relationship, and further that cows not avoiding humans in an approach test may even take some more pressure when handled in the milking parlour compared to the once that show longer AD to approaching humans.

Effect of a magnesium rich marine extract on welfare of growing pigs in response to acute stress

K. O'Driscoll[1], D. Lemos Teixeira[1], D. O'Gorman[2], S. Taylor[2] and L. Boyle[1]
[1]Teagasc, Pig Development Department, Animal and Grassland Research and Innovation Centre, Moorepark, Fermoy, Co. Cork, Ireland, [2]Celtic Sea Minerals, Currabinny, Carrigaline, Co. Cork, Ireland; keelin.odriscoll@teagasc.ie

The aim was to investigate if a magnesium rich marine extract (Acid-Buf™) would improve pig welfare in response to mixing and an out-of-feed event. 448 weaned (d28) piglets were assigned to Control (CL; Mg 0.16%) or Acid-Buf™ (AB; Mg 0.18%) diets in single sex groups of 14.7 pigs/pen were mixed with 7 of the same sex and diet from another pen: CL male, AB male, CL female and AB female (n=4 of each) at d56. At mixing, aggressive behaviour and no. pigs involved per bout was recorded for 3 hours from video. Postural behaviour was recorded at 10 min intervals. At d112 feed was removed for 21 hours. Pens were observed continuously for 8×2 min periods after re-introduction of the feed and aggressive behaviour was recorded. Skin lesions of 4 focal pigs/pen were scored on the day prior to, and after, mixing and the out of feed event. Saliva samples were collected on d56 and d113 (1 h before and 1, 3 and 8 h after mixing/feed delivery post deprivation) and at 10:00 on d55, d57 d58, d112 and d114 by allowing the 4 focal pigs to chew on a cotton bud for 1 min. Cortisol was analysed by ELISA. Data were analysed in SAS (Proc Mixed). At mixing, aggressive interactions lasted longer in males than in females (16:55 vs 34:27 mm:ss, s.e. 03:38; $P<0.01$) and more CL than AB pigs were involved/bout (2.13 ± 0.39 vs 2.08 ± 0.34; $P<0.05$). There was no effect of diet or sex on skin lesion scores, but AB females had lower cortisol concentrations than CL (1.51 ± 0.12 vs 1.91 ± 0.13 ng/ml; $P<0.05$). During the out-of-feed event, neither sex nor diet affected salivary cortisol levels, but males were more aggressive than females (0.182 vs 0.122 aggressive interactions/pig/min; s.e. 0.019; $P<0.05$), and CL pigs had higher skin lesion scores than AB pigs (13.2 ± 1.1 vs 10.0 ± 1.0; $P<0.05$). Supplementation with magnesium had some beneficial effects on pig welfare.

Analysis of owner/shipper certificates for transport of horses from the USA for slaughter in Canada

R.C. Roy and M.S. Cockram

University of Prince Edward Island, Sir James Dunn Animal Welfare Centre, Department of Health Management, Charlottetown, C1A 4P3, Canada; mcockram@upei.ca

Following the closure of equine slaughter plants in the USA, concern has been expressed over the welfare of horses transported from the USA for slaughter in Canada. This study aimed to provide quantitative information on the journeys experienced by horses transported from the USA to slaughter plants in Canada and thereby assess potential welfare issues. A Freedom of Information Act request was made to the USDA for all records of owner/shipper certificates for 2009. These certificates require the shipper to self-certify that they are in compliance with USDA regulations on the transport of equines for slaughter. On some certificates there was incomplete information and parts of the certificate were blacked out. Records for 1,800 loads, comprising 52,136 horses showed that the horses transported to the 6 federally approved equine slaughter plants operating in Canada in 2009 (2 in Alberta, 2 in Quebec, 1 in Ontario and 1 in Saskatchewan) were from 16 states in the northern part of the USA. Out of the 862 loads for which the type of facility for origin of journey was made available, 32% were from auction centres, 33% from feedlots and 35% from horse collection centres. Seventy-four percent of the horses were Quarter horses, 6% thoroughbreds, 5% draught horses, <1% ponies and the remainder were classified as miscellaneous. Fifty-nine percent were mares, 40% were geldings and 1% were stallions. The median number of horses in each load was 29, range 13 to 41. Horses were sent for slaughter throughout the year. Journey duration was calculated under the assumptions that the load was inspected at the time of unloading and that the certificate details were correct in that they represented the final section of the journey with no rest periods for feed and water. For long journey durations these assumptions were unlikely to have been correct. In many certificates, the date and time of loading, and inspection date and time were blocked. Based on 379 certificates, the median journey duration was 19 h, $Q_1=5$, $Q_3=27$. Thirty-six percent of horses were transported for <6 h, 11% for 6-18 h, 13% for 18-24 h, 25% for 24-36 h, 9% for 36-48 h and apparently 6% >48 h. Few certificates contained any additional information on fitness. Only 2 certificates reported dead-on-arrival horses. This analysis provided useful information on journey characteristics and enabled welfare assessments conducted at individual equine slaughter plants to be placed in a wider context. If the journey durations are correct, some journeys exceeded those specified in regulations and based on other research studies would likely put these horses at risk of negative welfare outcomes, such as dehydration.

Comparison of the fluctuating asymmetry of broody and non-broody hens

J.L. Campo, S.G. Dávila and M.G. Gil
I.N.I.A., Animal Genetics, Carretera La Coruña km 7, 28040 Madrid, Spain; jlcampo@inia.es

The purpose of this study was to analyse the difference between broody and non-broody hens in fluctuating asymmetry, a measure of developmental stability that has been reported to reflect well-being status and performance of an animal. There were 12 different breeds housed in 48 litter floor pens, 4 pens for each breed and 20 hens in each pen. A total of 24 broody hens were observed. Fifteen broody hens were observed in pens of white shell egg breeds (1 in each of 3, 4, 3, 1 and 4 pens of the Blue Andaluza, Black-barred Andaluza, Black Red Andaluza, Black Castellana and White-faced Spanish, respectively), 7 broody hens were observed in pens of tinted shell egg breeds (1 in each of 1, 2, 2, 1 and 1 pens of the Buff Prat, White Prat, Birchen Leonesa, Quail Castellana and Quail Silver Castellana, respectively), and 2 broody hens were observed in pens of brown shell egg breeds (1 in each of 1 pen of the Red-barred Vasca and Red Villafranquina). Broodiness was induced naturally (change in duration of natural light and temperature in June). Fluctuating asymmetry was measured at 52 wk of age in a total of 48 hens (24 broody and 24 non-broody hens). The broody/non-broody pairs were matched for breed, 1 broody and 1 non-broody hen of each breed in each pen with a broody hen. The non-broody hen of each pen was randomly chosen. Right (R) and left (L) sides of four morphological traits (middle-toe length, leg length, wing length and leg width) were measured. Relative fluctuating asymmetry $[2(R - L)/(R + L)]$ was used for all traits. A factorial design was used with the statistical model: $xijk = m + Gi + bj + pk(j) + Gbij + eijk$, where $xijk$ is the analysed fluctuating asymmetry, m is the overall mean, Gi is the effect of the group (broody vs non-broody hens), bj is the effect of breed ($j = 1...12$), $pk(j)$ is the effect of pen within breeds (the number of pens was unequal in the different breeds, k ranging from 1 to 4), $Gbij$ is the interaction and $eijk$ is the residual. Group was considered fixed effect, whereas breed and pen were considered to be random effects. Although there was a trend to increase fluctuating asymmetry in broody hens in comparison to non-broody hens, there were no significant differences for fluctuating asymmetry of leg length (1.09±0.14 vs 0.89±0.14), leg width (4.63±0.65 vs 3.84±0.65), toe length (2.22±0.32 vs 1.78±0.32), wing length (1.65±0.30 vs 1.30±0.30) and the combined fluctuating asymmetry value of the four traits (2.20±0.21 vs 2.15±0.21). It was concluded that hens with more asymmetrical morphological traits would not be more likely to get broody.

Poster 171

Behavioural changes in Chinchillas as a result of confinement

T. Tadich and V. Franchi
Universidad de Chile, Fomento de la Producción Animal, Santa Rosa 11735. Santiago, 8820000, Chile; tamaratadich@u.uchile.cl

The Chinchilla lanígera, rodent endemic from Chile, has been domesticated, selected, and bred, due to the high quality of its fur. Intensive production systems have resulted in changes in their behavioural repertoire and the appearance of stereotypies. Changes in the presentation of behaviours can allow us determine the degree of plasticity and variability of their time budget, and this as the presentation of stereotypies can be used as indicators of animal welfare. The aim of this study was to determine the presentation of fur-chewing and other stereotypies in captive chinchillas. Ten chinchillas from a pelt production system were studied. Five were known fur-chewers (FC), and 5 without the behaviour were used as control (C) group. Focal continuous sampling was used for 24 hours. A system with infrared cameras and a DVR was implemented. Chinchilla's were kept in their individual cages under normal husbandry conditions. Sampling was done between April and May with light cycles of 11 hours (between 07:00 and 18:00 h). For behavioural analysis the Observer XT 2011® was used. All behaviours were recorded, and an ethogram composed by 32 behaviours was developed. Within and between group differences in time allocation and behavioural frequencies were analysed with AOV, Kruskall-Wallis, Two sample T-test and Wilcoxon Rank Sum analysis respectively. A p value of $P<0.05$ was set for statistical differences. No difference in time budget's between groups was found, resting was the main state (61,75%), and bar-chewing the most frequent event (338.9 times/day on average). Diurnal and nocturnal dedication to behaviours was compared, no significant differences between groups were found. Group C showed a higher dedication to feeding and resting during the day, and an increased nocturnal dedication to stereotypies, self-directed, locomotor, and other behaviours ($P<0.05$), with no differences for grooming between day and night. The FC group had no differences in time dedicated to feeding and other behaviours, but increased resting during day, and increased nocturnal behaviours such as stereotypies, grooming, fur-chewing, locomotion and self-directed behaviours ($P<0.05$). Because of the implications on production fur-chewing was analysed as a separate stereotype. All chinchillas showed at least one stereotype (bar-chewing, backflipping). When Fur-chewing was analysed together with grooming a lost of a circadian pattern appeared in FC group. All chinchillas showed at least one stereotype, being a welfare concern. The conditions in which chinchilla's are kept for pelting purposes should be assessed in more detail to avoid mental suffering of these animals during the productive cycle, as the economic consequences of behaviours such as fur-chewing.

Aggressive behaviours of commercially housed growing pigs at different group density

S.J. Rhim[1], S.H. Son[2], H.S. Hwang[3] and J.K. Hong[4]

[1]Chung-Ang University, School of Bioresource and Bioscience, 72-1 Naeri, Ansung, 456-756, Korea, South, [2]Chung-Ang University, School of Bioresource and Bioscience, 72-1 Naeri, Ansung, 456-756, Korea, South, [3]Chung-Ang University, School of Bioresource and Bioscience, 72-1 Naeri, Ansung, 456-756, Korea, South, [4]National Institute of Animal Science, Swine Science Division, 9 Eoryongri, Cheonan, 331-801, Korea, South; sjrhim@cau.ac.kr

This study was conducted to clarify the aggressive behaviours of commercially housed growing pigs at different densities per pen. Thirty groups of pigs housed at low, medium, and high group density (5, 10 or 20 individuals per 6.0×6.0 m pen) were consecutively observed for 10 h on days 30, 90, and 180 with the aid of video technology. The frequency of vocalization was lower at low group density and higher at high group density on all investigated days. Pigs housed at high group density showed significantly more aggressive behaviour than those at low group density. The investigation reveals a higher level of aggression in older pigs and at high group density. It is concluded that group density is a major cause of the observed aggressive behaviour.

Environmental and genetic parameters for the tail length of South African Merinos

S.W.P. Cloete[1,2], A.J. Scholtz[1] and A.C.M. Kruger[1]
[1] *Western Cape Department of Agriculture, Directorate Animal Sciences: Elsenburg, Private Bag X1, Elsenburg 7607, South Africa,* [2] *University of Stellenbosch, Department of Animal Sciences, Private Bag X1, Matieland 7602, South Africa; schalkc@elsenburg.com*

The docking of lamb's tails is not permitted in some European countries. However, tail-docking is still commonly practiced in southern hemisphere sheep producing countries, mostly to combat breech blowfly strike. This practice may not be sustainable in the long term, as it may be challenged on ethical grounds. Alternatives therefore need to be considered. Data of 1,698 Merino lambs in South Africa were used to derive genetic parameters for tail length when docked at 21.9 (SD=3.8) days using REML methods. The overall mean for tail length amounted to 22.1 (2.7) cm. Birth weight was included as a linear covariate to adjust lamb tail length for differences in body size. Tail length was independent of the sex of the lamb, but it was affected by year of birth (2007 to 2013), selection line (lines divergently selected for number of lambs weaned, various crosses and backcrosses among these lines), the age of the dam of the lamb (2 to 7+ years) and birth type (single or multiple). Age of dam effects followed a typical U-shaped curve with tail length of young and old dams being shorter than that of progeny of ewes of intermediate ages. This pattern is also commonly found in early growth traits such as birth weight, pre-weaning weight and weaning weight, and may persist to yearling age. Least squares means (±SE) for the tails of multiple lambs were lower than those of singles (20.7±0.4 vs 22.3±0.4 cm respectively). The heritability of tail length amounted to 0.30±0.06 in a single-trait analysis, while tail length was also affected by a significant maternal permanent environmental effect (0.08±0.3). These parameters remained similar in a two-trait analysis with weaning weight. Covariance ratios between tail length and weaning weight amounted to 0.18±0.18 for the genetic correlation, 0.48±0.16 for the dam permanent environmental correlation, 0.22±0.05 for the environmental correlation and 0.24±0.03 for the phenotypic correlation. These results suggest that lamb tail length is heritable, and would potentially respond to genetic selection for shorter tails in the South African Merino genetic resource flock studied. Such selection may be done without prejudicing weaning weight, as suggested by the genetic correlation between tail length and weaning weight. However, the practical implications of such a strategy need consideration before being embarked upon, as results indicate that there may be a higher correlation on the maternal level. Data recording over a longer period and including dams and granddams with records of their own are needed to elucidate this issue.

Poster 174

Effects of piglet age on some indicators assessing pain at tail docking, with or without analgesia

V. Courboulay[1], M. Gillardeau[1], M.C. Meunier-Salaün[2] and A. Prunier[2]
[1]IFIP, Institut du Porc, BP 35104, 35651 Le Rheu cedex, France, [2]INRA, UMR1348 Pegase, 35590 Saint-Gilles, France; valerie.courboulay@ifip.asso.fr

Pig tails are often docked at an early age in order to prevent tail biting later on. According to farms, it is carried out in the first two days but may occur later, in combination with other management procedures. An experiment was conducted in order to evaluate the effects of age and analgesia on pain related to tail docking, using different types of indicators. Two day-old piglets were allocated within litter to tail docking after being treated with meloxicam (M, 0.4 mg/kg, IM) or placebo (P) or to sham docking (S) (n=29/treatment). Movements and intensity of vocalisations were recorded during tail docking. Postoperative behaviours (activity, posture, localisation and tail movements) were observed every 2 minutes for an hour at 3 periods: immediately after the procedure, 4 and 24 hours later, for 24 animals per treatment. Blood samples were collected on 18 other piglets per treatment, 30 minutes after the procedure, to measure plasma cortisol. Wound healing was assessed on day 1, day 3 and day 5. Piglets were weighed at birth, before the procedure and at weaning. A similar protocol was applied on piglets treated at 5 days of age. Average daily gain from birth to weaning did not differ between treatments ($P>0.1$) but tended to be higher in D2 than in D5 piglets ($P=0.1$). Cortisol was higher in P than in S pigs ($P<0.05$) with M pigs being intermediate. Intensity of vocalisation and body movements during procedures were higher in P and M than in S pigs ($P<0.05$). Cortisol level decreased whereas vocalizations and body movements during procedures increased from D2 to D5 regardless of treatment ($P<0.001$). None of the behaviours recorded after the procedure, except being under the heat lamp and tail movements, differed significantly between treatments. S piglets presented more ample tail movements and less trembling tails than P and M ($P<0.001$). Pain-related behaviours such as huddling up and trembling were more performed on D2 than on D5 whereas social isolation was more frequent on D5. Animals were more under the heat lamp on D2. These age-related differences are probably due to higher thermoregulation needs in younger piglets. Wound healing was not altered by analgesia. Overall, it can be concluded that (1) P piglets showed more evidence of pain than S piglets with M piglets being either intermediate or close to P piglets, and (2) many behavioural indicators showed age-related variations that did not seem specific to pain.

Improving dairy cow welfare by a stretched closing phase of the pulsation cycle during milking

F. Bluemel, P. Savary and M. Schick

Agroscope Taenikon, Institut für Nachhaltigkeitswissenschaften INH, Taenikon 1, 8356 Ettenhausen, Switzerland; pascal.savary@agroscope.admin.ch

Despite normed installations of milking machines, farmers experience milking problems, such as: restlessness during milking, knocking off milking units, milk-ejection disturbances and performance decreases. These physiological and behavioural parameters are indicative of poor welfare in dairy cows during milking. Particularly the movement of the liner, which is controlled by the pulsators and causes pressure on the teat, can be painful and thus cause stress in dairy cows. The objective of this study is to evaluate the effect of two pulsation cycle types differing in closing and closed phase regarding the effect on hind leg activity (HLA) and milk removal of cows. The experiment was performed at the Federal Research Station Agroscope, Tänikon, Switzerland. Treatment A differed from treatment B in pulsation chamber cycle closing phase e.g. c-phase and closed phase e.g. d-phase (treatment A: c-phase=70 ms, d-phase=330 ms; treatment B: c-phase=130 ms, d-phase=270 ms). 36 dairy cows were randomly confronted with treatment A and B during 12 morning milkings. Milk flow was measured by milk flow meters (LactoCorder, WMB AG, Balgach, Switzerland). HLA was recorded by three-dimensional accelerometers (RumiWatch, Itin + Hoch, Liestal, Switzerland). Lactation stage of cows was classified in three categories (early, mid, late). Lactation number was summed up in first lactation and further lactations. HLA was observed throughout different milk flow phases. Positions of pedometers were distinguished between hind leg facing milking pit and opposite milking pit. For statistical evaluation, the linear mixed effects model of Pinheiro and Bates was used. Peak flow rate (PFR) was significantly higher (1.04 kg/min) in treatment B than in treatment A ($F_{1.379}=27.48$, $P<0.001$). Additionally, lactation number had a significant influence ($F_{1.35}=4.80$, $P<0.04$). Increased lactation number led to higher PFR (1.24 kg/min). Cows confronted with treatment B showed a tendency to a higher total milk yield (TMY) ($F_{1.379}=3.45$, $P=0.06$) at 0.213 kg more TMY than when milked with treatment A. In this study, TMY decreased significantly with increasing lactation stage ($F_{2.35}=14.87$, $P<0.001$) and increased significantly with increasing lactation number ($F_{1.35}=15.72$, $P<0.001$). No significant differences in HLA could be found between the two treatments ($F_{1.351}=0.03$, $P=0.87$), different milk flow phases ($F_{4.351}=0.03$, $P=0.82$) and positions of pedometers ($F_{1.351}=0.07$, $P=0.79$). A stretched c-phase leads to a higher PFR and appears to increase TMY. Differences in HLA could not be determined and thus no influence in welfare was apparent.

Effect of certain farming practices on the behaviour and welfare of cattle

N. M'Hamdi[1], R. Bouraoui[2], C. Darej[1], M. Ben Larbi[1], M. Ben Hammouda[3] and L. Lanouar[4]
[1]Institut National Agronomique de Tunis(INAT), Animal Sciences, 43 Rue Charles Nicole, Cité Mahrajène Le Belvédère Tunis, 1082, Tunisia, [2]Ecole Supérieure d'Agriculture Mateur (ESA Mateur), Route de Tabarka Mateur, 7030, Tunisia, [3]Institution de la Recherché et de l'Enseignement Supérieur Agricole (IRESA), Alain Safari, 1000, Tunisia, [4]Higher Institute of Agronomy of Chott-Meriem, Animal Sciences, BP47, Sousse, 4042t, Tunisia; naceur_mhamdi@yahoo.fr

Animal welfare or animal well-being refers to the physical and mental health of animals. An animal in a poor state of welfare may suffer from discomfort, distress, or pain, which may compromise its ability to grow, survive, and produce or reproduce. However, some farming practices are routine painful procedures carried out on cattle to facilitate management. The pain caused by these procedures and its alleviation may be evaluated by monitoring behaviour and physiological responses. Physiologic (heart and respiratory rates) and behavioural indicators (movement, feeding behaviour and behaviour in herds) have been used to assess acute distress responses to potentially painful husbandry procedures. A study was conducted in four dairy farms in the North of Tunisia to assess the effect of some farming practices (dehorning, vaccination, blood sampling, trimming and rectal search) on the behavioural and physiological response of 20 cows, 20 heifers and 20 calves. The results showed a significant effect of type of intervention ($P<0.05$), age ($P<0.001$) and the farm ($P<0.05$) on the behaviour and the physiological response of animal. Dehorning and trimming are the most painful operations in cows; they affect the physiology and behaviour of the animal. Indeed, average heart rates range from 87 beats/min during dehorning to 102 beats/min during trimming. The sex of animal has a significant effect ($P<0.05$) on the welfare of animal. Results showed that cows and heifers express pain more than calves.

Slaughter of pregnant cattle: aspects related to ethics and animal welfare

K. Riehn[1], G. Domel[2], A. Einspanier[3], J. Gottschalk[3], G. Lochmann[3], G. Hildebrandt[4], J. Luy[5] and E. Lücker[6]

[1]Hamburg University of Applied Sciences, Faculty of Life Sciences, Ulmenliet 20, 21033 Hamburg, Germany, [2]Landratsamt Altenburger Land, Lindenaustraße 9, 04600 Altenburg, Germany, [3]University of Leipzig, Institute of Physiological Chemistry, An den Tierkliniken 1, 04103 Leipzig, Germany, [4]Freie Universität Berlin, Institute of Food Hygiene, Königsweg 69, 14163 Berlin, Germany, [5]Freie Universität Berlin, Institute of Animal Welfare and Behaviour, Oertzenweg 19b, 14163 Berlin, Germany, [6]University of Leipzig, Institute of Food Hygiene, An den Tierkliniken 1, 04103 Leipzig, Germany; katharina.riehn@haw-hamburg.de

The EU has among the world's highest standards of animal welfare and the safety of the food chain is indirectly affected by the welfare of animals, particularly those farmed for food production. The welfare of food producing animals depends largely on how they are managed by humans. Harmonised EU rules are in place covering a range of food safety- and welfare-affecting issues but a regulatory framework which governs the slaughter of pregnant farm animals is still missing. A need for action on this part has not been seen by the European Commission yet, because it was assumed that pregnant heifers are only slaughtered in exceptional cases. However, first own investigations show, that the proportion of pregnant heifers raised in different European member states amounted up to 10%. These results are very similar to those seen in the United States and Asia. In this context, the object of this study was the collection of data concerning the frequency of slaughter of pregnant heifers in different German abattoirs. For this purpose a questionnaire was sent to different German slaughterhouses. The questionnaire evaluates the proportion of pregnant animals in the abattoir as well as animal welfare relevant parameters during transport, stunning and slaughter and the fate of the foetuses and unborn calves. Feedbacks from 53 slaughterhouses could be gathered and evaluated. More than 50% of the participants reported that they slaughter pregnant heifers regularly. The maximum percentage share of pregnant animals on the total number of female cattle is approximately 15%. 9.6% of the female slaughter cattle were pregnant on average, the median was 7.1%. More than 90% of the affected animals were slaughtered during last two trimesters of pregnancy.

Smothering in UK free-range flocks: incidence, location, timing and management

J. Barrett[1], A.C. Rayner[2], R. Gill[3], T.H. Willings[4] and A. Bright[2]
[1]University of Oxford, Department of Zoology, South Parks Road, Oxford, OX1 3PS, United Kingdom, [2]FAI Farms Ltd, The Field Station, Wytham, Oxford, OX2 8QJ, United Kingdom, [3]The Lakes Free Range Company Ltd, Meg Bank, Stainton, Penrith, CA11 0EE, United Kingdom, [4]Noble Foods Ltd, The Moor, Bilsthorpe, Newark, NG22 8TS, United Kingdom; annie.rayner@faifarms.co.uk

Smothering in poultry is an economic and welfare-related concern. This study presents the first results of a questionnaire addressing the incidence, location, timing and management of smothering from Noble and The Lakes free-range farm managers (representing 35.0% of the UK free-range egg supply). In total, 206 questionnaires were returned; 162 from Noble and 44 from The Lakes, translating to a 50.0 and 100.0% response rate respectively. On average, the reported flock mortality due to smothering was low (<5%) and smothering incidences were infrequent (38.8% of farm managers did not report any smother incidence in the last flock). However, there was high variability between farms. Moreover, the majority of farm managers reported that over 50% of their flocks placed had been affected by smothering. The location and timing of smothering (excluding smothering in nest boxes) tended to be unpredictable and varied between farms, although some popular reduction measures were identified, for example walking birds more frequently. A follow up study will investigate the correlations between smothering, disease and other welfare problems and may shed further light on management solutions.

Welfare effect of outdoor areas for dairy cows: exercise pen or pasture?

L. Aanensen, S.M. Eilertsen and G.H.M. Jørgensen
Bioforsk Nord Tjøtta, Parkveien 1, 8860 Tjøtta, Norway; lise.aanensen@bioforsk.no

From January 1'st 2014 Norwegian animal welfare legislation requires access to pasture minimum 8 weeks/year for all dairy cows, incl. cows in loose housing. Dairy barns are often located far from pastures and the herds are large. Farmers and the dairy industry seek solutions for the technical and practical challenges. Can a simple exercise pen replace access to pasture? In 2013, we performed a pilot study to investigate how dairy cows use different outdoor areas in two commercial dairy farms with loose housing and automatic milking (AMS). Farm 1: Fifty dairy cows, DeLaval milking robot, guided cow traffic. 2.8 ha green pasture for 33 days. Farm 2: Fifty dairy cows, Lely milking robot, free cow traffic. 0.7 ha exercise pen in a small forest area for 15 days. Outdoor access during daytime. Registered weather type and number of cows outside. Within farms, we compared activity, behaviour and milk production on days with and without access to outdoor areas. Farm 1. In average 66% of the cows went outside when possible. Cows exhibited significantly higher activity on 'pasture-days'; standing/walking 80% of the time, lying 20% of the time, walking in average 275 steps/day (9 am-4 pm). 72.2% of all observations were spent grazing, only 4.7% of the observations were lying. On 'indoor-days', cows were standing/walking 37% of the time, lying 63% of the time and walking 70 steps (9 am-4 pm). Milk yield was however lower ($P<0.01$) and number of visits to the AMS fewer ($P<0.001$), on 'pasture-days'. Fewer cows went out on rainy/drizzly or bright sunny /hot days compared to cloudy days ($P<0.05$). Farm 2. In average 31% of the cows went outside when possible. Access to the outdoor area gave no significant effect on cow activity ($P=0.8$). Main outdoor observations; standing/walking with head up (43.2%) and lying (33.2%). Even though they only had access to a small forest area, eating counted for 18.3% of the observations. The number of cows outside was mainly controlled by the indoor feeding interval and not by the weather ($P=0.115$). Access to the exercise pen did not affect the daily milk yield, but there was an increase in number of visits to the AMS ($P=0.005$). In conclusion, access to outdoor areas, preferably pasture, is important for dairy cows. Cows are natural grazing animals and pasture give them the opportunity to perform synchronized grazing behaviour and increases the available area. Increased space gives subordinate individuals the opportunity to avoid social conflicts and increased access to resources (feed, milking robot, lying cubicle). Consequently, this have a positive effect on animal welfare. Exercise pens with mixed surface and a stimulating environment have the same positive effects, although not fulfilling the grazing needs.

Budget time of growing rabbits housed in bicellular cages or collective pens at two ages

E. Filiou[1], A. Trocino[2] and G. Xiccato[1]
[1]University of Padova, Department of Agronomy, Food, Natural Resources, Animal and Environment, viale dell'Università 16, 35020 Legnaro (PD), Italy, [2]University of Padova, Department of Comparative Biomedicine and Food Science, viale dell'Università 16, 35020 Legnaro (PD), Italy; eirini.filou@studenti.unipd.it

The present study aimed at assessing differences of budget time at 52 and 73 d in growing rabbits (456) housed in conventional bicellular cages (0.112 m², 2 rabbits/cage) or open-top collective pens (20-54 rabbits/pen). The effect of pen size (small, 1.68 vs large, 3.36 m²) and stocking density (12 vs 16 rabbits/m²) in collective pens was also evaluated. The data were firstly analysed by a mixed procedure with housing system (bicellular cages vs collective pens), age and their interaction as fixed effects, and time of the day as random effect. The data of collective pens were then used to test the effect of pen size, stocking density, age and their interactions as fixed effects and time as random effect. The rabbits spent less time in feeding (7.76 vs 10.9%) and allo-grooming (0.65 vs 1.58%) and more time in moving (0.81 vs 0.35%) and resting (82.1 vs 77.7%) in collective pens than in bicellular cages (P<0.01); besides, rabbits in collective pens exhibited running, rearing and hopping and did not show stereotypic behaviours. Significant interactions (P<0.001) were measured between housing system and age: from 52 to 73 d feeding time decreased only in rabbits in collective pens; time of allo-grooming increased and time of moving and sniffing decreased only in rabbits in bicellular cages; the reduction of resting time with stretched body was more pronounced in bicellular cages (12.7 to 0.87%) than in collective pens (9.32 to 3.61%). In collective systems, both pen size and stocking density had weak effects. Within collective systems, when the age increased, rabbits spent more time self-grooming, allo-grooming, sniffing, and running, exhibited more hops and reduced feeding and resting time (0.05<P<0.001). During resting, the stretched position was preferred (8.37 to 3.24%) by younger than older rabbits. Some significant interactions between stocking density and age (P<0.01) were measured: at 52 d feeding time was lower in pens at 12 rabbits/m² compared to those at 16 rabbits/m² (7.19 vs 9.49%), whereas differences disappeared at 73 d (6.60 vs 6.35%); at 52 d, self-grooming (4.67%) and allo-grooming times (0.33%) were similar while they differed at 73 d of age (7.81 vs 6.55% and 0.35 vs 1.11%, respectively). In conclusions, rabbits in collective pens displayed a more complete behavioural pattern, despite resting more, than rabbits in bicellular cages. Within collective systems, variations in pen size and stocking density exerted only weak effects on a small number of behavioural traits.

Poster 181

Development and validation of an improved cognitive bias task for broiler chickens

O.S. Iyasere[1], J.H. Guy[1], A.P. Beard[1] and M. Bateson[2]
[1]Newcastle University, School of Agriculture, Food and Rural Development, Newcastle upon Tyne, NE1 7RU, United Kingdom, [2]Newcastle University, Centre for Behaviour and Evolution, Institute of Neuroscience, Newcastle upon Tyne, NE2 4HH, United Kingdom; o.s.iyasere@ncl.ac.uk

In a cognitive bias task to assess the affective state of animals based on their interpretation of ambiguous stimuli, animals are initially trained to discriminate rewarded and unrewarded stimuli and then tested with intermediate stimuli. It is hypothesised that animals that judge the ambiguous stimuli as similar to the unrewarded/punished stimuli are in a worse affective state. The aims of this study were, first, to develop an improved cognitive bias task for broilers by introducing a punisher (air puff) predicted to speed up the rate of learning a spatial task, and second, to validate this task on broilers whose state has been manipulated to mimic chronic stress by feeding corticosterone. 15 birds (approx. 15 days of age) learnt to discriminate between rewarded and unrewarded locations within 3 days of training (2 sessions /day). A two-day break was introduced between the end of training and the cognitive bias task during which corticosterone (4 mg/kg BW) dissolved in dimethyl sulfoxide (DMSO) was fed via injected mealworms to 8 randomly selected birds, while 7 control birds had mealworms injected with DMSO for 7 days and the cognitive bias task took place on day 3-5 of treatment. On each day of the cognitive bias task, each bird was offered three ambiguous locations between the rewarded and unrewarded locations and d the latency to approach these locations was recorded. On the 7th day of treatment, birds were weighed, euthanized and their internal organs recovered, weighed and expressed relative to body weight. The latency to approach the rewarded and unrewarded locations was compared with a paired sample t-test. Latencies to approach ambiguous locations were adjusted with respect to the rewarded and unrewarded locations. Adjusted latency, relative spleen and liver weights were analysed using GLM with treatment as a factor. Correlation of cognitive bias and indicators of chronic stress were undertaken. After six sessions, birds showed discrimination between rewarded and unrewarded locations ($P<0.001$) and during the cognitive bias task corticosterone birds took longer to approach ambiguous locations than control birds ($P<0.05$). Corticosterone birds had reduced relative spleen weight but with an enhanced relative liver weight ($P<0.05$). Relative liver weight was correlated with mean latency to approach ambiguous locations ($P<0.05$). In conclusion, the improved cognitive bias task enhanced learning in broilers. Treatment with corticosterone produced anatomical changes consistent with chronic stress and also made birds judge ambiguous locations more pessimistically.

Author index

Hansen, C.F.	226	Huuki, H.	136, 137, 194
Hansen, I.	189	Hwang, H.S.	251
Hansen, S.W.	103	Hyndman, T.H.	109, 141
Hansted, H.	165		
Harms, J.	150	**I**	
Haskell, M.J.	59, 91	Illmann, G.	214, 218
Haun Poulsen, P.	182	Ipema, B.	143
Hawkins, P.	83	Ishida, M.	94
Heath, C.A.E.	59, 144	Ito, S.	94
Heathcote, E.	72	Iyasere, O.S.	260
Heeres, J.	202		
Heise, H.	190, 212	**J**	
Held, S.	34	Jamieson, J.	245
Hemsworth, L.	69, 122	Jansens, J.	217
Hendrycks, W.	243	Jensen, K.K.	162
Henriksen, B.I.F.	98, 103	Jensen, M.B.	221
Hensel, O.	111	Jensen, T.B.	200, 209
Heres, L.	145	Jinman, J.C.	92
Hermans, K.	242	Jochemsen, H.	82
Hernández, H.	108	Johnson, A.	62
Herskin, M.S.	130, 221	Jørgensen, G.H.M.	240, 258
Heyerhoff-Zaffino, J.	49	Jorgensen, M.	222
Hietaoja, J.	244	Julliand, V.	107
Hildebrandt, G.	256	Junge, W.	150, 156
Hillmann, E.	192	Jurkovich, V.	101, 176, 224
Hindle, V.	44, 70		
Hoar, B.R.	138	**K**	
Hochereau, F.	41, 205	Kaarlenkaski, T.	42
Hockenhull, J.	90, 211	Karnholz, C.	169
Hofstede, G.J.	58	Karriker, L.	62
Holinger, M.	182	Kashiha, M.	56
Holmes, D.	182	Kasuya, E.	94
Holm Parby, L.	209	Kauppinen, T.	244
Hong, J.K.	251	Keeling, L.J.	121
Hopkins, J.E.	74	Kelton, D.F.	49, 87
Hopster, H.	67	Keppler, C.	63, 71
Horseman, S.	90	Kettlewell, P.J.	170
Hothersall, B.	179	Kézér, L.	101, 224
Hötzel, M.J.	193, 215	Kiessling, A.	77
Houdelier, C.	217	Kilbride, A.L.	61
Houe, H.	112, 161, 200, 221	Kjær Nielsen, H.	188
Houlton, S.	208	Klaffenböck, M.	125
Hovinen, M.	175	Kling-Eveillard, F.	40, 41
Huertas, S.M.	81, 84	Knierim, U.	50, 71
Hulsen, J.	160	Kniese, C.	113
Hunniford, M.	232	Knop, D.	182
Hurme, T.	175	Knowles, T.G.	90

Printed in the United States
by Baker & Taylor Publisher Services